CW00722992

Sustainable Management of Sediment Resources, Volume 4

Sustainable Management of Sediment Resources

Sediment Management at the River Basin Scale

The European Sediment Research Network SedNet was funded as a Thematic Network project (contract No. EVK1-CT2001-20002) by the 5th European Framework Programme for RTD, under the Key Action "Sustainable Management and Quality of Water" of the Environment Programme, topic 1.4.1 Abatement of Water Pollution from Contaminated Land, Landfills and Sediments.
SedNet website: www.SedNet.org

Sustainable Management of Sediment Resources

Sediment Management at the River Basin Scale

EDITED BY

DR. PHILIP N. OWENS

National Soil Resources Institute, Cranfield University, North Wyke Research Station, Okehampton, United Kingdom.
Now at: Environmental Science Program, University of Northern British Columbia, Prince George, Canada

Amsterdam • Boston • Heidelberg • London • New York • Oxford
Paris • San Diego • San Francisco • Singapore • Sydney • Tokyo

Elsevier
Radarweg 29, PO Box 211, 1000 AE Amsterdam, The Netherlands
Linacre House, Jordan Hill, Oxford OX2 8DP, UK

First edition 2008

Library of Congress Cataloging-in-Publication Data
A catalog record for this book is available from the Library of Congress

British Library Cataloguing in Publication Data
A catalogue record for this book is available from the British Library

ISBN: 978-0-444-51961-0
ISBN-set: 0-444-51959-9
ISSN: 1872-1990

For information on all Elsevier publications
visit our website at books.elsevier.com

Printed and bound in Hungary

08 09 10 11 12 10 9 8 7 6 5 4 3 2 1

Introductory comments by UNESCO IHP

Erosion and sedimentation processes and management in catchments, river systems and reservoirs have reached global importance. The socio-economic and environmental impacts of erosion and sedimentation processes in river basin management are significant. Regrettably, it is estimated that over 50% of the world's reservoir storage capacity could be lost due to sedimentation within the next few decades. The situation is particularly severe in most of the developing countries. Accordingly, sediment management practices should be improved; even though various sediment transport models are at our disposal today, the inadequacy of knowledge about sediment production processes hinders practical progress in addressing problem-solving. The issue calls for integrated solutions where land-use management and water management are not decoupled.

The International Sediment Initiative (ISI) has been launched by UNESCO as a major activity of the International Hydrological Programme (IHP), with the aim to support the solving of sediment-related problems. IHP created ISI as a vehicle to foster international cooperation in handling regional sediment problems and in identifying local solutions. ISI also intends to promote international information exchange and to provide direct access to policy makers in Member States while encouraging scientific and professional communities in all regions and countries concerned. Focus was initially brought into the realization of a first Global Evaluation of Sediment Transport (GEST), the setting up of a global erosion and sediment information system and the review of erosion and sediment-related research worldwide. The initiative now aims to implement case studies for river basins as demonstration projects and, of course, educational and capacity-building efforts for sustainable sediment management.

ISI is open to collaboration with all interested institutions, international, regional or national associations and networks. As far as Europe is concerned, SedNet is certainly one of the first and most enthusiastic partners to have joined in and developed with us a fruitful cooperation. One of the results is the creation of a Danube Working Group. It involves the International Commission for the Protection of the Danube River (ICPDR), SedNet and IHP partners in the region that committed to draft and support a roadmap towards advice on the implementation of sediment management in the Danube WFD River Basin Management Plan. The annual SedNet conference hosted by the UNESCO Office in Venice in November 2006 laid the ground for a series of catchment-oriented roundtable discussions – one of which was devoted to the Danube with concrete proposals to be further developed in 2007–2008. One proposal supported by the UNESCO Office in Venice will lead to the establishment of a first sediment balance of the Danube River by the end of 2008.

UNESCO IHP very warmly welcomes the publication, within the SedNet series of books, of the present work dedicated to "Sediment Management at the River Basin Scale", including a contribution from ICPDR and ISI on the Danube River, the most complex and challenging river basin in Europe. There is no doubt that this publication will be essential reading to all those concerned with sediment-related issues within the larger framework of Integrated River Basin Management in Europe.

I would like, therefore, to express my warmest thanks to SedNet, all authors and all those who have provided the necessary support to guarantee the success of this publication. UNESCO IHP, for its part, shall do its best to facilitate and promote its dissemination among the widest audience possible.

Andras Szollosi-Nagy
Deputy Assistant Director-General for Natural Sciences
Secretary of the International Hydrological Programme of UNESCO

Edited by Philip N. Owens

Preface

This book represents one of four in a series by the European Sediment Network (SedNet), published by Elsevier, entitled *Sustainable Management of Sediment Resources*. The titles and editors of the books in this series are:

- *Sediment Quality and Impact Assessment of Pollutants*, edited by Damia Barceló and Mira Petrovic;
- *Sediment and Dredged Material Treatment*, edited by Giuseppe Bortone and Leonardo Palumbo;
- *Sediment Risk Management and Communication*, edited by Susanne Heise; and
- *Sediment Management at the River Basin Scale*, edited by Phil Owens.

The history behind these books is interesting and helps to explain their content and focus. The European Sediment Research Network (SedNet) was a European Commission (EC)-funded Thematic Network project (contract number EVK1-CT2001-20002) within the 5th European Framework Programme (FP5), within the Key Action theme "Sustainable Management and Quality of Water" of the Environment Programme, topic 1.4.1 "Abatement of Water Pollution from Contaminated Land, Landfills and Sediments". The SedNet project arose from a call by numerous scientists and stakeholders for a network aimed at bringing people together to discuss and review sediment issues within European river systems. It also recognized that there was a need for a state-of-the-art review of sediment management issues and how they related to EU policy, such as the Water Framework Directive (WFD), so that appropriate sediment management guidance could both assist with the implementation of such policy and could help to shape future policy development. The SedNet project was thus born and was funded for three years (2002–2004) by the EC. Given its original mandate and role within topic 1.4.1, attention mainly focused on sediment within European river basins (as defined by the WFD), and thus largely neglected estuarine and marine sediment, and focused on contaminated sediment or processes relevant for the management and abatement of contaminated sediment. The themes and contents of the four SedNet books reflect this focus, although efforts have been made to provide a wider context.

Between 2002 and 2004, the main activities of SedNet were 17 workshops and three conferences, a regular newsletter, a website (www.sednet.org), and reports and documents on sediment management in European river basins: see the SedNet website for details of these. SedNet activities were originally organized around a series of working groups (WGs) which were guided by the SedNet coordinator, Jos Brils, and a stakeholder panel made up of representatives of interested organizations such as Hamburg Port Authority, Hamburg-Harburg Technical University, TNO, Port of Rotterdam, UNESCO and Venice Port Authority. Each of the WGs had a "leader" and a core group. In the case of WG4, the core group comprised:

- Sabine Apitz (UK):
- Ramon Batalla (Spain);
- Alison Collins (UK);
- Marc Eisma (The Netherlands);
- Heinz Glindemann (Germany);
- Sjoerd Hoornstra (The Netherlands);
- Harald Köthe (Germany);
- Phil Owens (Leader, UK);
- John Quinton (UK);
- Kevin Taylor (UK);
- Bernhard Westrich (Germany):
- Sue White (UK); and
- Helen Wilkinson (UK).

WG4 organized a series of workshops:

- Existing guidelines and the EU Framework Directives, Silsoe, UK, 28–29 October 2002;
- Sources and transfers of sediment and contaminants in river basins, Hamburg, Germany, 26–28 May 2003;
- Modelling and other decision support tools for sediment management, Lleida, Spain, 10–11 November 2003; and
- Societal cost benefit analysis and sediments, Warsaw, Poland, 18–19 March 2004.

It also produced reports and statements based on these workshops (see the SedNet website). In addition, some of the information was directly or indirectly published in journal and book papers (e.g. [1–5]) and contributed to more general SedNet publications [6].

During the mid-term review of SedNet in Brussels in 2003, a series of recommendations were put forward, which shaped the form and activities of SedNet for the second half of the three-year EC-funded phase. One of these recommendations was that the pre-existing working groups were to be restructured into five work packages (WP):

1. Coordination, synthesis, dissemination and stakeholders panel;
2. Sediment management at the river basin scale;
3. Sediment quality and impact assessment;
4. Sediment and dredged material treatment; and
5. Sediment risk management and communication.

Thus WG4 became WP2 (here on referred to as WP2), with a revised theme.

A second important recommendation was that each of the four main WPs (WP2–WP5) produce a book, as part of a series, that reviewed much of the activities of the workshops and also reviewed other relevant material, and tried to offer appropriate suggestions for the sustainable management of sediment that were relevant to the WP theme. This book represents one of those, and is the effort of SedNet WP2. Given the time between the initiation of the idea of the book series (2003) and their final publication in 2007, inevitably there has been much change in the group working on this particular book. Some members of the WP2 core group have changed jobs or roles and consequently have had to withdraw from becoming actively involved with the production of manuscripts, although all have been involved with ideas and material that have been incorporated within the chapters. In some cases, other representatives have been found to lead and co-author the chapters. Thus the chapters in the book represent the work of not only the authors and co-authors, but also the other members of the WP2 core group, and the organizers and participants of the four WP2 workshops. In addition, ideas have also developed from interaction with the other WPs and also the SedNet stakeholders panel.

SedNet is now in a new phase. Funding from the EC finished at the end of 2004 and since this time SedNet has developed into a stand-alone network (renamed the European Sediment Network, SedNet) and is self-financed through contributions from the organizations which represent the steering group. Most of its initial objectives still remain, but it is also a much wider-ranging network and addresses all sediment issues and environments, including estuarine and marine environments. It continues to organize workshops and conferences, and produce documents and publications (e.g. [7]), many of which are described or published in the *Journal of Soils and Sediments*.

There are many people to whom thanks are due. A special thanks goes to all those involved with WP2 and/or this book over the last 6 years, many of whom are listed above and/or are authors of chapters in this book, and to the SedNet "family". It has been a privilege to interact with such an outstanding group of scientists and stakeholders, who have been dedicated to furthering the appreciation and recognition of sediment in Europe. I am also extremely grateful to Alison Collins for her help with the first phase of SedNet, and to Alison Foskett for her help with the production of this book. The book was produced as camera-ready-copy, using a template supplied by the publishers, and this required considerable effort and dedication from all those involved – it is hoped that readers are sympathetic towards any errors or problems of chapter layout. Thanks are extended to Piet den Besten, Ulrich Förstner and Wim Salomons who acted as internal SedNet referees of the chapters in this book, to David Kenyon who produced the index, to Joan Anuels and Andrew Gent of Elsevier for their patience and understanding, and to Jürgen Busing who, as the EC scientific officer responsible for the first phase of SedNet, encouraged us all the way. Perhaps the greatest thank you goes to Jos Brils for his incredible motivation and enthusiasm for all things to do with sediment.

These four books represent an important contribution to the literature on sediment dynamics and management in European river basins. While a huge amount of effort and time has been put into them, they only represent a start. Although the focus necessarily has been largely restricted to Europe, many of the sediment processes and issues are of relevance to other parts of the world. It is hoped, therefore, that this book, and the others in the SedNet series, encourage further activities to advance our understanding, appreciation and management of sediment in European countries and throughout the world.

References

1. Köthe H (2003). Existing sediment management guidelines: an overview. Journal of Soils and Sediments, 3: 144–162.
2. Owens PN, Apitz S, Batalla R, Collins A, Eisma M, Glindemann H, Hoornstra S, Köthe H, Quinton J, Taylor K, Westrich B, White S, Wilkinson H (2004). Sediment management at the river basin scale: synthesis of SedNet Working Group 2. Journal of Soils and Sediments, 4: 219–222.
3. Gerrits L, Edelenbos J (2004). Management of sediments through stakeholder involvement. Journal of Soils and Sediments, 4: 239–246.
4. Owens PN (2005). Conceptual models and budgets for sediment management at the river basin scale. Journal of Soils and Sediments, 5: 201–212.
5. Owens PN, Batalla RJ, Collins AJ, Gomez B, Hicks DM, Horowitz AJ, Kondolf GM, Marden M, Page MJ, Peacock DH, Petticrew EL, Salomons W, Trustrum NA (2005). Fine-grained sediment in river systems: environmental significance and management issues. River Research and Applications, 21: 693–717.
6. Salomons W, Brils J (Eds) (2004). Contaminated Sediment in European River Basins. European Sediment Research Network, SedNet. TNO Den Helder, The Netherlands.
7. SedNet (2007). Sediment Management – An Essential Part of River Basin Management Plans. SedNet, The Netherlands.

Phil Owens
Prince George, Canada, May 2007

Contents

Sustainable Management of Sediment Resources: Sediment Management at the River Basin Scale
Edited by Philip N. Owens

Sediment Behaviour, Functions and Management in River Basins

Philip N. Owens

National Soil Resources Institute, Cranfield University, North Wyke Research Station, Okehampton, Devon EX20 2SB, UK. Now at: Environmental Science Program, University of Northern British Columbia, 3333 University Way, Prince George, British Columbia, V2N 4Z9, Canada

1. Introduction

The purpose of this chapter is to set the scene for the other chapters in this book and to place these chapters within a wider context of sediment and its management. It presents some concepts on what sediment is and what it is composed of. Assessment of sources and pathways, and estimates of sediment fluxes at different spatial and temporal scales, are described so as to illustrate the nature and magnitude of sediment behaviour and dynamics. Some of the main functions and uses of sediment, and the natural and anthropogenic influences and impacts on these, are also described. These considerations naturally lead to an assessment of how to manage sediment so as to balance the needs of nature and society, and to a discussion on the river basin as an appropriate management unit to do this.

2. What is sediment?

Sediment means different things to different people and consequently there are a variety of different terms and phrases used to describe 'sediment'. 'Mud', 'dirt' and 'sludge' are terms that are often used by the public or non-scientific community when referring to 'sediment', although mud is also a term used by certain groups of scientists when referring to fine-grained organic and inorganic material (i.e. clay- and silt-sized material), as opposed to coarse-grained 'sediment'. For many, especially managers and regulators, sediment is synonymous with dredged material. It is perhaps here that some of the problems and issues of sediment management arise, i.e. the lack of appreciation and agreement on what sediment is.

In terms of definitions of sediment, there are several (see Table 1 in Chapter 3, this book). A useful definition is that put forward by the European Sediment Network, SedNet (www.sednet.org):

Sediment is suspended or deposited solids, of mineral as well as organic material, acting as a main component of a matrix which has been or is susceptible to being transported by water [1, 2].

This definition is what many would regard as appropriate. However, it is not fully inclusive, and does fail to recognize other forms of transportation such as wind and ice (e.g. glaciers), and indeed it can be argued that sediment movement by people, animals, machinery, etc. is relevant. Also, sediment need not necessarily move in suspension. Large sediment particles may move by rolling, saltation or sliding. This helps to highlight the problem of understanding and defining what is meant by sediment. Thus the definition stated above is a good working definition, but it is important to bear in mind the caveats just described.

It is useful to consider the different components and forms of sediment, and how these affect the behaviour, function and management of sediment. Material that is in solid form is usually distinguished from that which is in solution. A boundary between the two is often set at 0.45 μm, although this is an arbitrarily defined boundary determined by laboratory analytical procedures. Sediment is thus usually defined as material >0.45 μm. Colloidal material is frequently ignored, but represents ultra-fine particles, usually within the range 0.001–1 μm. While colloidal material may represent a relatively small proportion of transported or deposited sediment, it is likely to be important in terms of contaminant transport.

Sediment particles are either mineral or organic. The former are denser than the latter and have different hydrodynamic, physical and chemical properties. Finer sediment particles, such as clay-, silt- and fine sand-sized particles, are usually transported and deposited as aggregates or flocculated material (for a discussion on the difference between the two types, see Droppo *et al.* [3]). These aggregates or flocs consist of four main components: mineral particles, organic material, water and air (Figure 1).

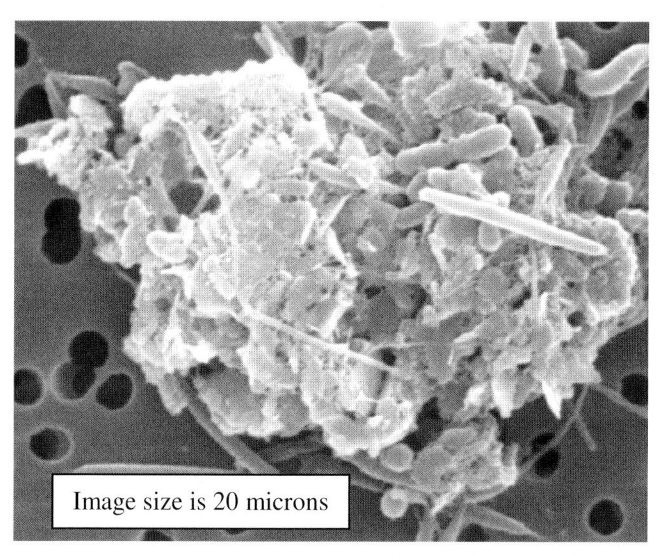

Figure 1. A scanning electron microscope image of a sediment floc, which is a composite particle composed of organic and inorganic sediment particles and voids containing water and air. The dark circles are holes within the filter paper (photo: D.J. Arkinstall, reproduced with permission)

It is the relative composition of these four main components that determine the behaviour and properties of the flocculated sediment (Figure 2). In particular, the composition of aggregated or flocculated material affects its density and thus the settling characteristics of suspended sediment. They also control the forces required to resuspend sediment deposited on the channel bed, such as the surficial fine-grained laminae [4]. Recent research (e.g. [5–7]) has highlighted the importance of organic matter, and in particular the colloidal particles of extracellular polymeric substances (EPS) manifested as fibrillar material, on the structure and transport behaviour of flocs. Research has also demonstrated the compartmentalization of contaminants *within* flocculated sediments [8].

While much of the scientific community recognizes the importance of flocculation in (re)defining what constitutes fine-grained, cohesive, suspended sediment [5], much of the management community still consider that fine-grained sediment moves as individual, discrete particles.

Larger sediment particles, such as coarser sands, gravels, etc., are mainly transported as individual, discrete particles, although they may have surface coatings of finer sediment particles such as clays. Table 1 presents a typical classification of sediment based on particle size.

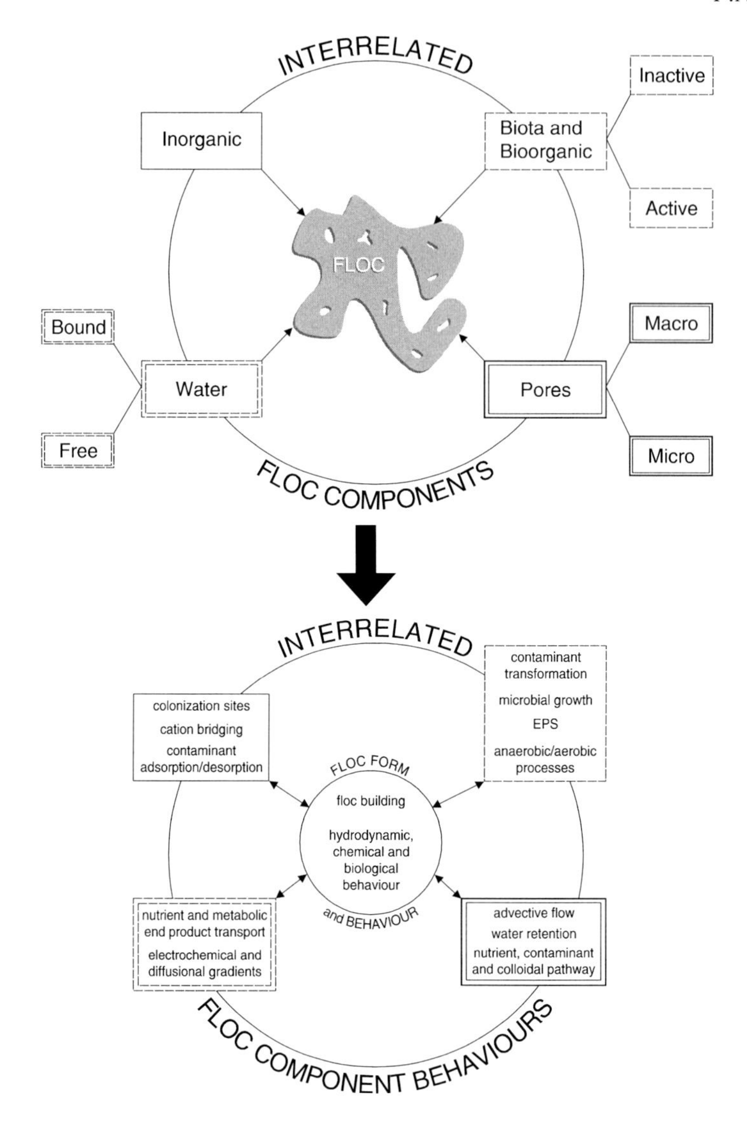

Figure 2. Conceptual model of the interrelationship between form and behaviour of a sediment floc. The model shows the linkages between the individual components and their behaviour on overall floc form and behaviour (source: [5] reproduced with the permission of Wiley)

Table 1. Particle size classification for sediment particles (source: adapted from [9])

Class (Wentworth)		Diameter size (mm)	Size (phi scale)
Boulder	Very large	4096–2048	−12 to −11
	Large	2048–1024	−11 to −10
	Medium	1024–512	−10 to −9
	Small	512–256	−9 to −8
Cobble	Large	256–128	−8 to −7
	Small	128–64	−7 to −6
Gravel	Very coarse	64–32	−6 to −5
	Coarse	32–16	−5 to −4
	Medium	16–8	−4 to −3
	Fine	8–4	−3 to −2
	Very fine	4–2	−2 to −1
Sand	Very coarse	2–1	−1 to 0
	Coarse	1–0.5	0–1
	Medium	0.5–0.25	1–2
	Fine	0.25–0.125	2–3
	Very fine	0.125–0.0625	3–4
Silt	Very coarse	0.0625–0.0312	4–5
	Coarse	0.0312–0.0156	5–6
	Medium	0.0156–0.0078	6–7
	Fine	0.0078–0.0039	7–8
	Very fine	0.0039–0.0020	8–9
Clay	Coarse	0.0020–0.0010	9–10
	Medium	0.0010–0.0005	10–11
	Fine	0.0005–0.0002	11–12
Colloidal		0.0010–0.000001	

3. Sediment movement and behaviour

In river catchments, sediment can be transported by a variety of mechanisms including: flowing water; wind; gravity-driven processes such as mass movements and bank collapse; flowing glaciers and ice; animals and humans; and machinery (such as tractors). In perennial river channels, sediment transportation is by flowing water (i.e. the river), but it is important to recognize that other processes are important outside the channel and that these processes can supply sediment to the channel. Thus, wind processes may be important in mobilizing and transporting sediment from exposed soil on fields or fine material stored as talus on hillslopes towards river channels. Wind and flowing water are important for transporting and delivering fine sediment (i.e. clay-, silt- and sand-sized material) from land to rivers, but the sediment load of a river also consists of coarser material such as gravels and boulders (Table 1). This coarser component is delivered to the channel by, for example, mass movements (such as landslides, rockfalls and debris flows) and the collapse of channel banks, and these processes may or may not involve flowing water.

Thus there are many different sources of sediment in river basins, and different mechanisms and pathways by which they are delivered from the source to the river channel, and these are described in more detail in Chapter 4 (this book). In addition to the fluvial sources of sediment in river basins, in the downstream, estuarine and near-coastal parts of a 'river basin' sediment is also supplied from estuarine, coastal and marine sources. In many cases, these non-fluvial sources may be dominant for the downstream parts of the basin. Thus in the Humber estuary, UK, about 97% of the sediment supplied to the estuary (ca. 6×10^6 t $year^{-1}$) is from coastal and marine sources, mainly the erosion of coastal cliffs and from the North Sea [10]. It is estimated [10] that only about 3% is derived from fluvial sources. Coastal and marine sediment sources are also important for other river systems, such as the River Elbe, and have important implications for how sediment is managed in the lower reaches of the River Elbe, including Hamburg harbour.

Within aquatic systems there is usually a simple distinction between the suspended load and bedload. The former is essentially sediment that is transported suspended within the water column and typically consists of material <2 mm in diameter (i.e. between 0.45 μm and 2 mm). The latter is that portion which moves by rolling, sliding and saltation and is therefore usually transported close to the channel bed. Bedload material is coarser and/or denser than the suspended load, the former being typically >2 mm in size, and has different hydrodynamic and chemico-physical properties than the finer, suspended load. There are more complicated classifications of the sediment load of a river with, for example, divisions of the suspended load into washload and suspended bed-material load components [9]. For simplicity, however, a separation into suspended sediment and bedload is usually sufficient, although it is important to recognize that the distinction between the two loads is time and space dependent, as material transported as bedload during one event may be transported in suspension during another event with greater flow velocity.

4. Sediment fluxes and storage

4.1. Sediment concentrations and fluxes

There have been numerous studies that have estimated sediment fluxes (sediment mass transported past a specific location per unit time, i.e. t $year^{-1}$) in river basins, over a range of temporal and spatial scales (Table 2). Most studies have been concerned with fluxes over relatively short periods of time, such as during high-flow events and over periods of a year or years [11], often as part of river monitoring programmes (see also [12, 13]). Sediment flux data rarely span more than a few decades at best, although there are records extending back for

Table 2. Some of the main temporal and spatial scales relevant for sediment fluxes (source: adapted from [15, 16])

Space	km^2	Time	Years
Global	10^{8-9}	Geological	10^8
Regional/multi-national	10^{6-8}	Quaternary	10^6
River basin	10^{2-4}	Holocene	10^4
Reach	10^{0-1}	Recent/historical	10^2
Particle	$10^{<-2}$	Present/annual	10^0
		Event	10^{-2}

about 100 years or so for some rivers [14]. Most sediment is transported during high discharge events such as those caused by precipitation, snowmelt (e.g. freshets), and water released from dams (natural and artificial impoundments). There are also situations when high sediment fluxes in rivers are not related to variations in water flows in rivers, and in these situations sediment is delivered to river channels from landslides and other mass movements, channel bank collapse, or anthropogenic disturbances such as mining and dredging activities.

Suspended sediment concentrations in flowing water vary by orders of magnitude, from essentially zero at low flow conditions (i.e. base flow) to >10 g l^{-1} during peak transport conditions (i.e. storm events and freshets) in some flowing water systems. In other systems, such as lowland chalk rivers, suspended sediment concentrations may always be relatively low, i.e. <100 mg l^{-1}. Values during extreme events, such as volcanic eruptions and glacial lake outburst floods, can be even greater: such events probably also result in the highest specific sediment yields (sediment mass transported to a specific location, per unit contributing area per unit time, i.e. t km^{-2} $year^{-1}$), although the occurrence and duration of such transport events are relatively limited [17]. Similarly, bedload fluxes range from essentially zero for most of the time to values over 10 kg s^{-1} m^{-1} during high-magnitude events [18]. Although sediment fluxes are generally greatest from highly disturbed agricultural and deforested basins, sediment fluxes and yields from urban basins can also be high, possibly >500 t km^{-2} $year^{-1}$ in some urbanized basins [19].

At the global scale, the flux of sediment to the oceans (defined here as a collective term for all oceans and seas, which are ultimately connected) has been estimated to range from <10 to >50×10^9 t $year^{-1}$. Most of these values are for suspended sediment, due to the problems associated with measuring bedload transport [13], although suspended sediment is believed to dominate the flux of sediment in the middle and downstream reaches of most rivers. Unfortunately, few of the values of flux presented in the literature (such as those contained in Table 3) have estimates of uncertainty or errors associated with them, and thus emphasis should be placed on the order of magnitude of these values rather than the specific numbers. The present consensus is that the global flux of (fine-grained) sediment to the oceans is of the order of 15–20×10^9 t $year^{-1}$.

Table 3. Some of the existing estimates of sediment flux from land to the global ocean (based mainly on suspended sediment fluxes) (source: adapted from [14, 31, 32], see these papers for references)

Study/author	Mean annual sediment flux ($\times 10^9$ t year^{-1})
Fournier (1960)	51.1
Kuenen (1950)	32.5
Gilluly (1955)	31.7
Jansen and Painter (1974)	26.7
Pechinov (1959)	24.2
Syvitski (1992)	24.0
Schumm (1963)	20.5
Milliman and Syvitski (1992)	20.0
Goldberg (1976)	18.6
Holeman (1968)	18.3
Syvitski *et al.* (2005)	16.2/12.6[a]
Milliman (1991)	16.0
USSR Nat. Comm. for the IHD (1974), Alekseev and Lisitcina (1974)	15.7
Dedkov and Gusarov (2006)	15.5
Sundborg (1973), Walling and Webb (1983)	15.0
Lvovitch *et al.* (1991)	14.9
Ludwig and Probst (1996)	14.8
Stallard (1998)	14.0
Milliman and Meade (1983)	13.5
Lopatin (1952)	12.7
Mackenzie and Garrels (1996)	8.3
Corbel (1964)	5.2

[a]First and second values excludes and includes the impact of dams, respectively.

It is estimated that the transfer of sediment by rivers from the land to the oceans accounts for about 95% of the sediment entering the global ocean [20]. The values of global sediment flux presented in Table 3 must be evaluated in light of estimates of the sediment being mobilized from the land by soil erosion by water processes, which are believed to be of the order of 50–75×10^9 t year^{-1} [14]. If other sediment sources were to be taken into account (see Chapter 4, this book), then the amount of sediment delivered to rivers would be much higher. The difference between estimates of sediment yield derived by soil erosion and values for sediment delivery to the global ocean reflects intermediate storage effects, such as in floodplains, lakes, ponds, reservoirs, etc. (see [21–23] and Chapter 4, this book). Indeed, some estimate that the annual storage of sediment upstream of large dams and impoundments is between 5 and 25×10^9 t year^{-1} [14, 24, 25], and Syvitski *et al.* [26] estimate that >100×10^9 t of sediment have been sequestered behind human-made reservoirs. Although the balance between sediment mobilization, conveyance and delivery is complex, and changes over both time and space [14, 22, 27], it is likely that for most river basins more sediment is in intermediate storage zones than is delivered to the lower reaches of rivers and ultimately the global ocean. Over long time periods, however, sediment flux to the global ocean will approximately equal the sum of sediment flux from all contributing sources.

The relationship between sediment mobilization from sources and downstream sediment delivery is often expressed as the sediment delivery ratio [28, 29]. Recently, the usefulness of this concept has been questioned by Parsons *et al.* [30] because of: (1) problems relating sediment fluxes and loads to those specific parts of a basin that are contributing sediment (termed contributing source areas), because some relatively small areas of the basin may contribute most of the sediment and other areas may contribute no sediment; and (2) issues of quantifying sediment transit distances and transit times. Parsons *et al.* [30] advocate that linear length of slope or channel may be a better measure of contributing source area than the entire upstream catchment area.

Spatially, sediment fluxes vary considerably throughout the world. The highest fluxes are usually associated with mountainous areas [26, 33], especially those experiencing tectonic activity and/or anthropogenic disturbances such as deforestation. Most of the sediment transported to the global ocean is by rivers in southeast Asia (Figure 3).

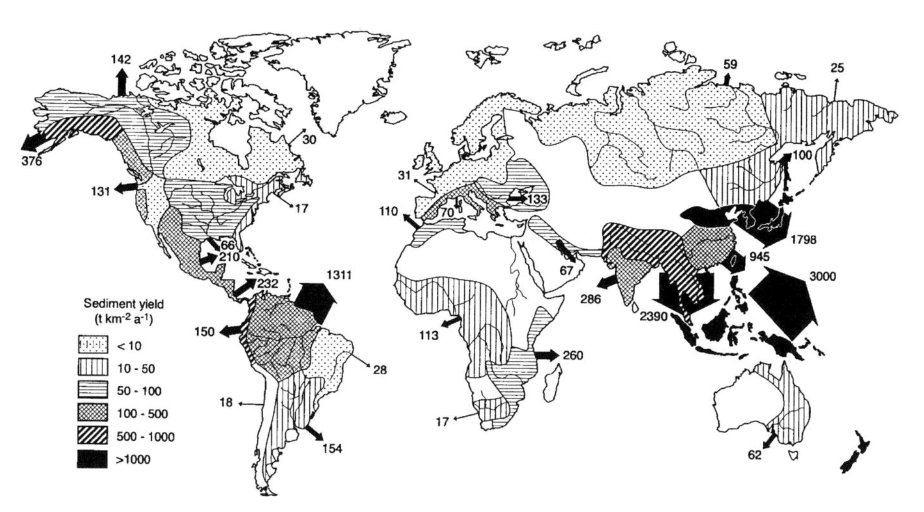

Figure 3. Annual fluvial sediment yields and sediment flux from large drainage basins to the oceans. Arrow sizes are proportional to sediment flux (in 10^6 t year^{-1}, shown as numbers) (source: from [34] based on [35], reproduced with the permission of Elsevier)

From the world map (Figure 3), it is seen that sediment yields and fluxes in the downstream reaches of European river basins are generally low, with the highest rates recorded for Mediterranean and mountainous river basins. Based on the FAO database of sediment fluxes in the downstream reaches of major rivers [36], the sediment flux from the European land mass to receiving oceans and seas is of the order of between 290 and 390×10^6 t year^{-1}, depending on whether Russia is included (giving a total contributing area of 9.8×10^6 km^2) or excluded (contributing area of 4.9×10^6 km^2). Thus the range in average specific sediment yield (SSY) is ca. 40–60 t km^{-2} year^{-1}. Data from the EUROSION project [37] gives a similar estimate of ca. 320×10^6 t year^{-1} from a contributing area of 4.6×10^6 km^2, giving an average SSY of ca. 70 t km^{-2} year^{-1}. These values differ from estimates by Owens and Batalla [38] and Syvitski *et al.* [26]. The latter estimate a modern suspended sediment flux of 680×10^6 t year^{-1} from a contributing area of ca. 10×10^6 km^2, giving a SSY of ca. 70 t km^{-2} year^{-1}, based on detailed sediment flux data for major rivers and extrapolations using models. Owens and Batalla [38] constructed a sediment budget for Europe based on estimates of sediment generation and sediment loss due to storage and removal (Figure 4) and estimate that the flux of sediment to the lower reaches of rivers and discharged into oceans and seas could be 714×10^6 t year^{-1} (from a contributing area of 6×10^6 km^2), giving a SSY of ca. 120 t km^{-2} year^{-1}. It is important to note that the downstream sediment flux estimate of Owens and Batalla [38] also includes the sediment which is deposited in lowland zones (such as harbours, estuaries and deltas).

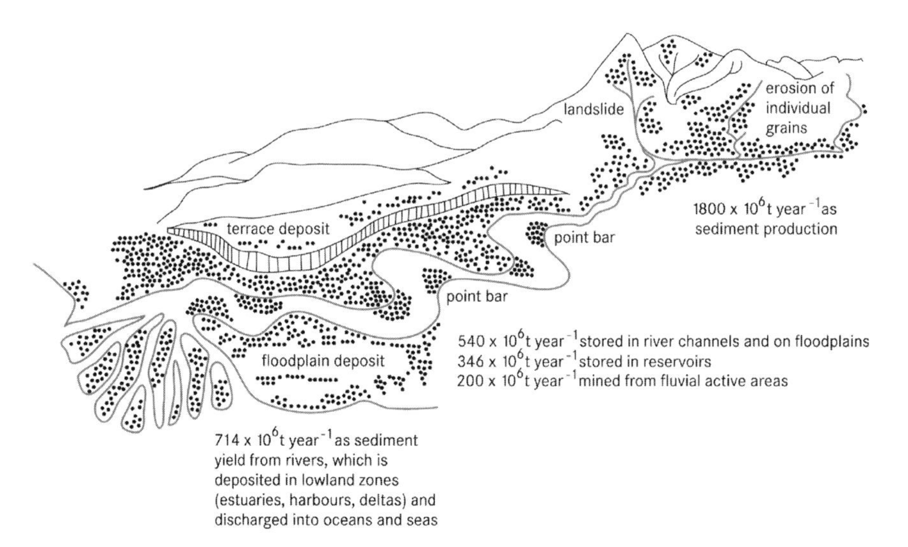

Figure 4. An approximate sediment budget for Europe (source: from [22], modified from [38], reproduced with the permission of Ecomed Publishers)

Sediment fluxes vary temporally as well as spatially (Table 2) in response to various natural and anthropogenic driving forces. Examples of the former include tectonic activity, weather (e.g. precipitation) and climate changes (glacial–interglacial as well as present global climate changes). Examples of anthropogenic changes include changes in land use and management, and river use and management [14, 27, 31], and in particular dam construction. Some examples of recent (i.e. events to decades) changes in sediment fluxes due to natural causes and anthropogenic activities are given in Chapter 4 (this book). There have also been changes in fluxes over longer time periods, and Figure 5 gives examples for various rivers.

To illustrate the effects of society on sediment fluxes, Syvitski *et al.* [26] estimated that the global land to ocean flux of suspended sediment and bedload before human influences were 14 and 1.5×10^9 t year^{-1}, respectively. These values compare to their contemporary estimates of suspended sediment flux of 16.2×10^9 t year^{-1} (or 12.6×10^9 t year^{-1} if the effects of reservoirs in trapping sediment in upstream locations are taken into account) and bedload flux of 1.6×10^9 t year^{-1}. Dedkov and Gusarov [32], however, determined that the 'natural' and 'anthropogenic' components of the present day global sediment flux are ca. 6 and 10×10^9 t year^{-1}, respectively. Over even longer time periods, Panin [31] estimates that the sediment flux to the global ocean ranged between 2.7 and 5.2×10^9 t year^{-1} between the late Jurassic and the Pliocene.

Although there is a general consensus that sediment delivery to rivers is increasing [26], mainly due to land use changes, there are important spatial variations in how fluxes and yields in rivers are changing, with values in some rivers increasing, while in others values are decreasing in recent decades (Figure 6). For Europe, there seems to be a general trend of values increasing in the west and Mediterranean parts of Europe, and values decreasing or stable in eastern Europe. Importantly, sediment fluxes seem to be increasing in many of the developing parts of the world, such as countries in South America, west Africa and South-East Asia (Figure 6).

Using a different dataset, Syvitski *et al.* [26] present a spatial classification of changes in sediment fluxes from land based on a variety of criteria including continent, climatic zone and elevation (Table 4). Table 4 shows an increase in sediment fluxes from the tropical climatic zone and a decrease from the temperate climatic zone. The importance of mountainous lands as the source of much of the sediment is also clearly shown. Interestingly, the absolute contributions from lowlands and the coastal plain are relatively small, and while there has been an increase from these lands over time they are fairly low.

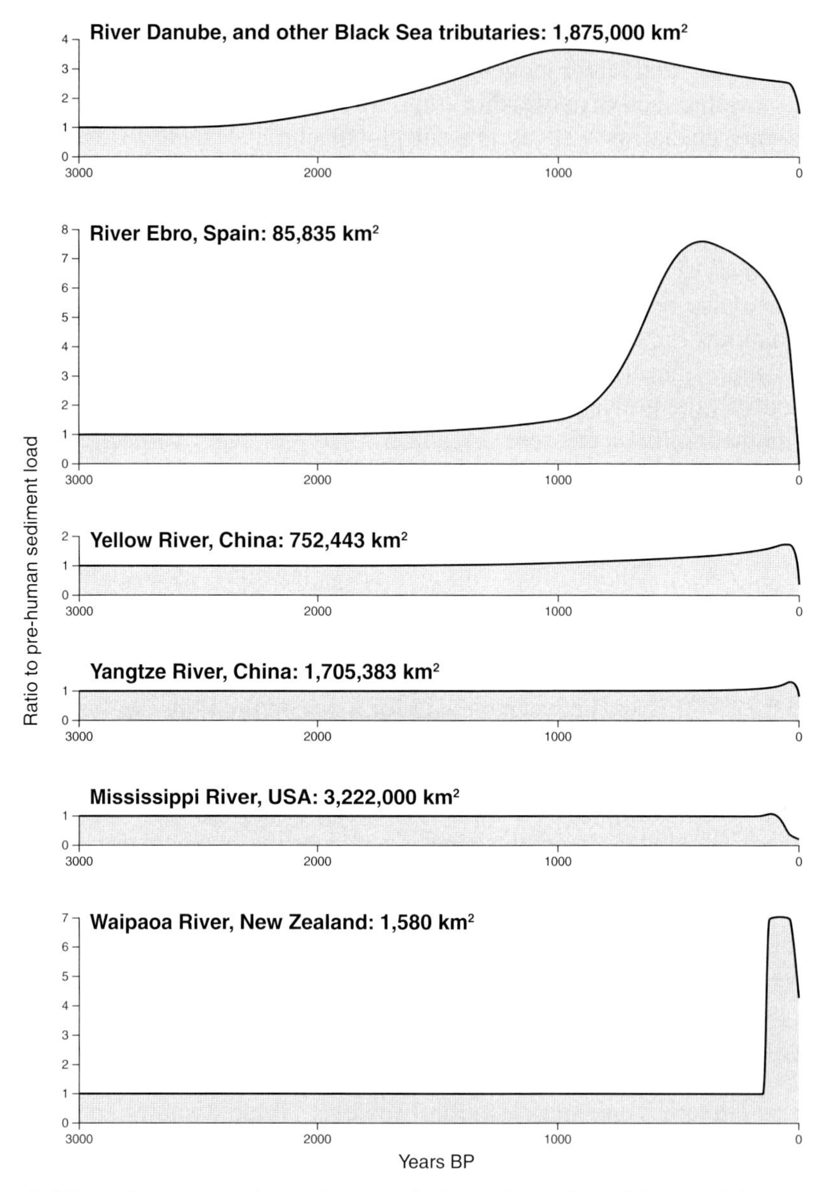

Figure 5. Schematic reconstructions of changes in the sediment flux of six world rivers over the past 3000 years. Examples illustrate: the complex response of rivers to changes in land use and increased supply of sediment (e.g. the Danube River (started ca. 2000 years ago), the Waipaoa River (started ca. 120 years ago)); sediment reduction in downstream reaches due to the effects of dams in trapping sediment between source and delivery to oceans (e.g. the Ebro River); and the effects of intermediate storage which tends to buffer fluvial systems to change thereby resulting in relatively stable sediment fluxes (e.g. Yellow River, Yangtze River and Mississippi River) (source: [14], reproduced with the permission of Elsevier)

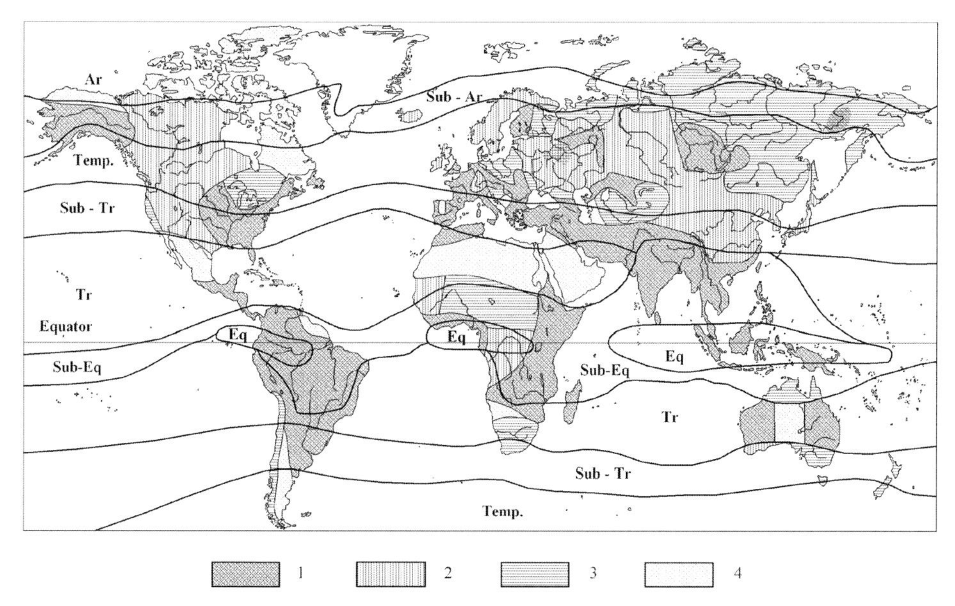

Figure 6. Trends of erosion intensity and suspended sediment yield changes during the second half of the twentieth century: 1, increasing; 2, decreasing; 3, relatively constant; 4, no data; Ar, arctic; Tr, tropical; Eq, equatorial; Temp, temperate (source: modified from [32], reproduced with the permission of the authors)

5. Sediment functions and anthropogenic impacts on these

5.1. Sediment functions

Sediment and its movement through river basins from source to sink are important for several reasons, some of which have already been alluded to above, and these include [14, 39]:

- as part of the global denudation cycle;
- for global biogeochemical (including carbon) cycling;
- for transferring nutrients and contaminants from terrestrial to freshwater to marine and coastal systems;
- for being (i.e. sediment itself) and creating (e.g. beaches, channel islands, saltmarshes) aquatic habitats and landforms;
- by helping to maintain a high level of biodiversity within aquatic systems through the creation of diverse sedimentary environments;

- for providing an important natural resource (e.g. aggregates, fertile soil on floodplains); and
- for the functioning of coastal ecosystems and the evolution of deltas and other coastal landforms.

Some of these functions are discussed in Salomons and Brils [1], Heise [2], Salomons [40] and the chapters in this book, and the reader is directed to these.

In recent years, the link between sediment (amount, type and dynamics) and the ecology of aquatic systems has become increasingly important, with various ecological metrics often used as an integrated measure of the health of a system. Studies (e.g. [41]) have shown that biotic assemblages in rivers may be influenced by sediment amount and composition. Conversely, other studies (e.g. [42]) have shown how in-stream vegetation influence sediment deposition, thereby illustrating the important inter-relationship between aquatic biota and

Table 4. Predictions of the seasonal flux of sediment to the world's coastal zone under modern and pre-human conditions (source: modified from [26])

		Area ($\times 10^6$ km^2)	Sediment flux ($\times 10^9$ t year^{-1})		Seasonal %			
			Prehuman	Modern	1[a]	2	3	4
Landmass	Africa	20	1.31	0.80	30	28	22	20
	Asia	31	5.45	4.74	8	12	49	31
	Australasia	4	0.42	0.39	26	27	26	21
	Europe	10	0.92	0.68	29	40	18	13
	Indonesia	3	0.90	1.63	31	28	19	21
	N. America	21	2.35	1.91	15	24	33	28
	Ocean Islands	0.01	0.004	0.008	25	13	38	25
	S. America	17	2.68	2.45	21	32	29	18
Ocean	Arctic	17	0.58	0.42	2	20	63	15
	Atlantic	42	3.85	3.41	20	30	27	23
	Indian	15	3.81	3.29	12	12	46	30
	Inland seas	5	0.47	0.14	13	51	28	8
	Med. and Black Seas	8	0.89	0.48	43	42	9	7
	Pacific	18	4.43	4.87	18	23	33	26
Climate	Tropical	17	1.69	2.22	22	17	29	32
	Warm temperate	47	9.07	8.03	18	22	35	25
	Cold temperate	17	1.94	1.46	17	35	30	19
	Polar	24	1.33	0.90	2	24	58	17
Elevation	High mountain	21	5.12	4.10	11	18	44	27
	Mountain	30	2.97	2.19	20	28	31	21
	Low mountain	36	4.67	4.80	20	23	31	25
	Upland	10	0.91	1.06	24	24	28	23
	Lowland	8	0.33	0.36	21	34	26	19
	Coastal plain	1	0.03	0.10	27	40	20	13
Global		106	14.0	12.6	18	23	35	25

[a]1 is December–February, 2 is March–May, 3 is June–August, 4 is September–November.

sediment dynamics. As an example of some of the various ecological zones associated with sediment at the sub-reach scale of a river, Figure 7 shows the interaction between sediment and water within a step-pool bedform. The variations in sediment particle size and structure provide important habitats for different types of aquatic life (see also Figure 9). This variation in sedimentary habitats at different spatial scales is important for maintaining biodiversity within aquatic systems by providing suitable conditions for spawning, shelter, food sources, etc.

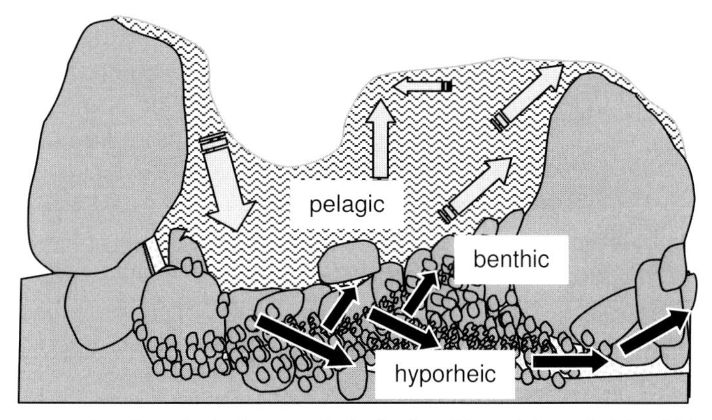

Figure 7. An example of ecological zones (pelagic, benthic and hyporheic) within a step-pool bedform. Grey arrows show general water flow above the bed and black arrows show potential flow through the substrate (source: from [43] reproduced with permission of the author)

5.2. *Impacts on sediment functions and dynamics*

Society is, however, manipulating the landscape for agriculture, industry, transportation and recreational use, amongst other things, and these are having important impacts on sediment functions and uses, and most of these impacts can be considered as detrimental. It is well documented, for example, that land clearance for agriculture has increased rates of soil erosion and sediment fluxes in rivers [14, 23, 26, 27]. In turn, this has increased the pressure on system functioning, for example, either through the removal of important topsoil at a rate faster than its formation, or through excessive fine-grained sediment amounts in rivers altering water temperature, reducing light infiltration into the water column or smothering important habitats like fish spawning gravels [39, 44–47]. Increasingly, the fisheries literature is realizing the importance of sediment for fisheries ecology and management, both as a key part in the formation of the habitat (e.g. sediment creating the channel bed substrate), but also in terms of the deleterious effects associated with excessive fine-grained sediment [46, 47]. Newcombe and MacDonald [44] and Petticrew and Rex [48] describe that salmonid fisheries can be affected by inert sediment:

- acting directly on free-living fish, either by killing them or by reducing their growth rate or resistance to disease, or both;
- interfering with the development of eggs and larvae;
- modifying natural movements and migrations of fish; and
- reducing the efficiency of methods used to catch fish.

Table 5 presents examples of the effects of suspended sediment on salmonid fisheries. Importantly, it is not just the concentration of suspended sediment that influences fish mortality but also the duration of exposure to high concentrations, which when combined is expressed as the stress index.

Table 5. Examples of exposures to suspended sediment that resulted in lethal responses in salmonid fisheries (source: modified from [44], see this paper for references listed below)

Species	Sediment (C, mg l^{-1})	Duration (D, hour)	Stress index ($\log_e$ [C×D])	Mortality effect (%)	Reference
Arctic grayling	25	24	6.4	6 – sac fry	Reynolds *et al.* (1988)
	23	48	7.0	14 – sac fry	
	65	24	7.4	15 – sac fry	
	22	72	7.4	15 – sac fry	
	20	96	7.6	13 – sac fry	
	143	48	8.8	26 – sac fry	
	185	72	9.5	41 – sac fry	
	230	96	10.0	46 – sac fry	
Chinock salmon	488	96	10.8	50 – smolts	Stober *et al.* (1981)
Coho salmon	509	96	10.8	50 – smolts	
	1217	96	11.7	50 – pre-smolts	
	18,672	96	14.4	50 – pre-smolts	
Chinock and sockeye salmon	207,000	1	12.2	100 – juveniles	Newcombe and Flagg (1983)
	9,400	36	12.7	50 – juveniles	
Rainbow trout	200	24	8.5	5 – fry	Herbert and Richards (1983)
	200	168	10.4	8 – fry	
Whitefish	16,613	96	14.3	50 – juveniles	Lawrence and Scherer (1974)

While many anthropogenic activities on the land have tended to increase rates of soil erosion and sediment delivery to river channels, in many parts of the world sediment fluxes in river channels have actually decreased in recent decades due to the construction of large dams and impoundments for hydro-electric power generation, irrigation and flood control, or small farm ponds for water supply and nutrient management. The trapping of sediment in reservoirs and ponds has also had detrimental effects on rivers through a reduction in downstream sediment fluxes, which in turn has impacted on aquatic habitats, such as saltmarshes and deltas, and by causing channel downcutting and the undermining of bridges and other infrastructure [49]. In-stream gravel mining has also reduced downstream sediment fluxes and resulted in similar detrimental effects [50]. Such impacts are discussed in more detail in Chapter 4 (this book).

6. Managing sediment in river basins

From a societal point of view, sediment is managed in the landscape for a variety of reasons, including:

- to maintain urban drainage and sewerage systems;
- 'maintenance dredging' in river channels, estuaries, ports, harbours, etc. to maintain shipping transportation;
- to maintain the life-span of reservoirs and for operational reasons;
- to ensure the efficient flow of water in watercourses and reduce flooding;
- to maintain or improve terrestrial and aquatic habitats (such as fisheries, coral reefs, etc.);
- to maintain geomorphological features, sometimes for aesthetic or recreational needs (such as gravel bars, beaches, etc.); and
- to maintain or improve water quality.

Consequently, there are a variety of different influences and impacts on sediment within a river basin, and therefore reasons why sediment is managed, and many of these are illustrated in Figure 8.

What becomes clear, when we look at (a) the multitude of functions and uses of sediment and (b) those factors that influence and impact on these functions and uses, is that: there are many functions and influencing factors; the interactions between them are complex; and they operate at different spatial locations within a river basin and operate at different time scales. The influence of sediment on flooding and floodplain development is, for example, usually relevant in the middle and downstream reaches of rivers, while sediment influences on reservoir life-span and operations are more relevant in the upper

Figure 8. Schematic representation of some of the main influences and impacts on sediment within a river basin, also illustrating the various functions that sediment performs, and the various stakeholders interested in sediment and its management (source: from [51] reproduced with the permission of Ecomed Publishers)

and middle reaches, although the downstream effects of dams, such as undermining of bridges, etc., will impact the lower reaches. Similarly, contaminated sediment will influence water quality for decades and perhaps centuries if there are considerable quantities of contaminated sediment stored on floodplains and on the channel bed of the river (i.e. historical or legacy contamination), while sedimentation on roads and highways is often a short-lived management issue.

An interesting, and topical, example illustrating sediment functions, and how society and nature are influencing these, is the Mississippi River, USA. The wetlands and marshes in the lower reaches of the river, along the Louisiana coastline, are controlled by sediment fluxes from the contributing basin. The wetlands are home to numerous species of birds, fish and trees. Over the last century, 400 km^2 of marshland have been destroyed, in part due to the construction of navigation channels but also as a result of a reduction of sediment supply to these marshes due to the construction of bank levees which have reduced overbank flooding. Currently there is a debate [52, 53] as to how to restore the Louisiana coastline and wetlands, which includes consideration of how to allow some of the sediment load of the river, of 120×10^6 t $year^{-1}$, to reach the wetlands again. One option is to split the Mississippi River into two, so that it flows into the Gulf of Mexico either side of the Birdsfoot delta, which could cost billions of US dollars and affect shipping routes [52]. Some, however, believe that such an approach is flawed, because hurricanes represent the main mechanism by which sediment is delivered to the wetlands (i.e. from sediment in the near coastal zone and not the river basin) [53]. The on-going debate illustrates an important ecological function that sediment has and highlights the difficulty in coming to a management solution that balances the needs of nature and society in an acceptable way.

Thus, while we tend to think of each sediment issue in relative isolation, and manage these accordingly, each sediment function or use is both dependent on other functions in time and space, and in turn influences many other sediment functions and uses. Thus if we are to manage sediment for the needs of nature (i.e. for maintaining fish habitats) and/or society (e.g. dredging for maintaining navigation), then this needs to be undertaken with a full appreciation and consideration of management impacts on nature and society within the river basin. Thus the river basin scale represents the most convenient and meaningful management unit for river management, be it for water and/or sediment.

7. Why manage sediment at the river basin scale?

Following on from the discussion in the previous section, there are several reasons, many of which are inter-related, as to why sediment management should either be at the river basin scale or be part of a broader management programme at this scale (i.e. for soil–sediment–water management). The following sections consider some of these and are based on Owens [22].

7.1. Interventions have implications

Decision-making needs to be placed within the context of a river basin because a local or site-specific intervention will in most cases impact other parts of the basin, either upstream or downstream of the intervention. This is because a river basin operates and functions as an open system with interconnected subsystems (hillslopes, floodplains, river channels, lakes, harbours, etc.). By altering one subsystem or part of a subsystem there will be impacts on other parts of the system. Thus, in order to manage the system in a sustainable way, this needs to be done at the most appropriate scale. For rivers, the management scale of the system is the river basin scale because the size and topography of the river basin, and the activities within it, control the sources, pathways and fluxes of water, sediment and contaminants. Figure 9 illustrates the interconnectivity of the channel system at different spatial scales nested within a river basin, and helps to show the variations in sedimentary structures at these scales. This variation is also important for maintaining aquatic biodiversity (see Figure 7).

7.2. Multiple functions, uses and users of sediment

As described above (also see Chapter 4, this book), most river basins throughout the world are highly populated and/or modified by human activities, and thus society has many uses of sediment and/or has various impacts on sediment behaviour which place pressure on the various functions that sediment performs (see Figure 8). Thus site-specific sediment intervention or management will have impacts on other functions, uses and users of sediment. It is therefore necessary to consider, and to some extent evaluate, all users and uses of sediment within a river basin (see Chapters 6 and 7, this book). The river basin scale is the most appropriate scale for decision-making involving multiple interested parties because the basin topography defines the area in which most sediment functions operate and in which many sediment users reside. Thus the actions of a farmer or land owner will influence soil erosion

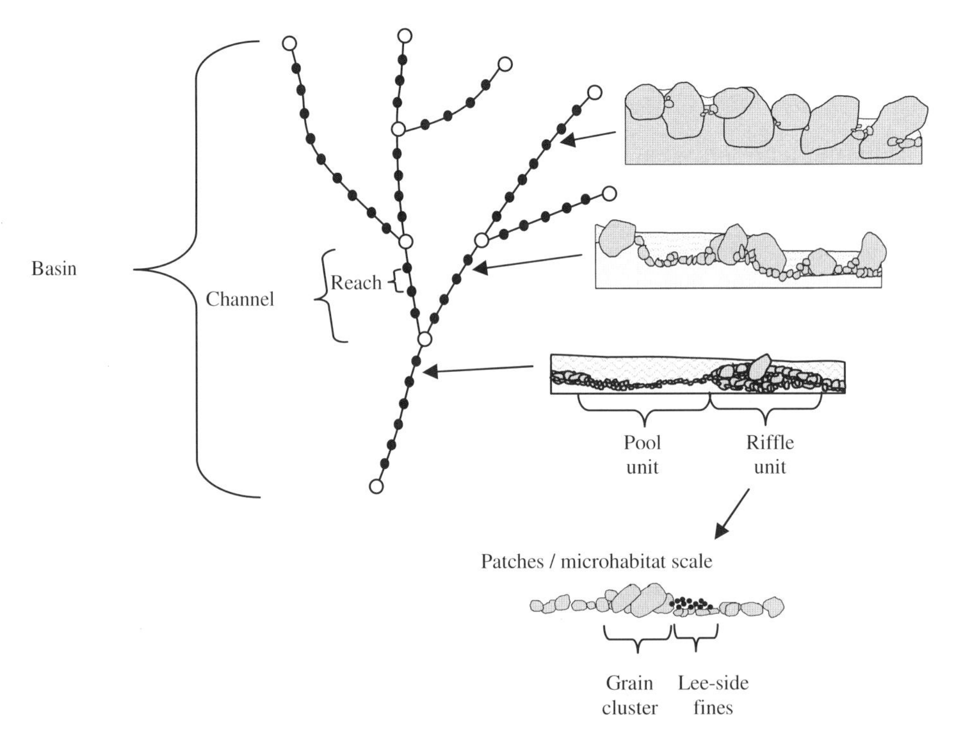

Figure 9. Conceptual diagram illustrating the different morpho-sedimentary environments at different settings and spatial scales within a river basin. Open circles identify nodes or confluences and solid circles series of reaches within a channel. The diagram illustrates the interconnectedness of these aquatic environments within the river basin, which also help to maintain biodiversity (source: modified from [43], reproduced with the permission of the author)

and sediment delivery within the basin in which the land is located, and thus downstream sediment functions and uses such as fishing and dredging, but their actions are unlikely to influence such functions and uses in adjacent basins.

7.3. Source control as the best solution

In most cases, source control will be the optimal long-term solution: environmentally, socially and economically. Most sources of sediment, and many sources of contaminants, are derived from diffuse sources. Most diffuse sources of sediment operate across large areas and may be dispersed throughout all or most of the river basin, such as those sources associated with agricultural land. The controlling of such diffuse sources necessitates a river basin scale approach in order to: identify all or most of the sources of the sediment and contaminants; for conducting meaningful risk assessment and evaluation; and to be able to implement remediation and mitigation options that are appropriate for controlling diffuse sources spread over a large area.

A good example to illustrate the last point is the Illinois River, USA, a tributary of the Mississippi River. Demissie *et al.* [54, 55] report that sedimentation and associated water-quality problems have been a major issue for several decades and have resulted in a joint federal and state programme, the Conservation Reserve Enhancement Program (CREP), in 1998 to manage sediment at the river basin scale, and in particular to encourage conservation measures to reduce sediment delivery and fluxes. An important part of CREP is the establishment of a monitoring programme and a sediment budget at the river basin scale to identify the major sources and pathways of the sediment (Figure 10), and also as a tool to assess the progress of the conservation programme over time by continuing to monitor these sources, especially hot-spots, for the next decade or so.

7.4. The dual issues of quantity and quality

Recently, in many countries, sediment management has had to consider the dual issues of sediment quantity and sediment quality [56]. The latter has become particularly important in recent years due to the introduction of guidelines and legislation associated with the removal and disposal of contaminated sediment, especially in marine and estuarine environments. In particular, the introduction of the EC Water Framework Directive (WFD, 2000/06/EC) now requires that issues of water quality and ecological status are addressed within a set timeframe. Of fundamental importance for water, and indirectly sediment, management is the focus on the river basin as the main unit of assessment and management, and the development of River Basin Management Plans (for further details see Chapters 2 and 3, this book).

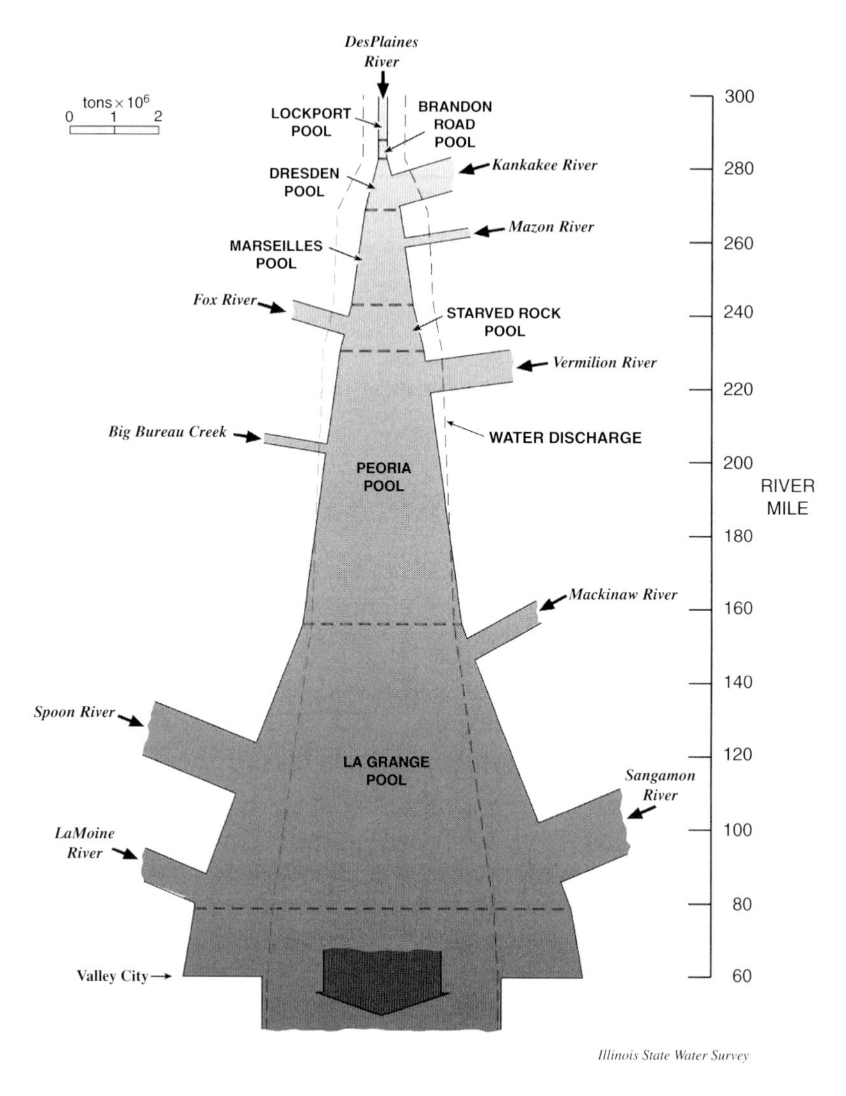

Figure 10. Sediment budget for the Illinois River, USA, for the period 1981–2000. Note to convert river miles to river kilometres multiply by 1.61, and to convert tons to metric tons divide by 1.10 (source: from [54, 55], reproduced with the permission of the authors)

8. What information do we need to manage sediment at the river basin scale?

Having established that the river basin scale represents the most appropriate scale or unit for management, it is necessary to obtain the relevant information required to make decisions so as to manage sediment effectively and, ideally, sustainably. This need, and associated recommendations, are considered in a series of four books by the European Sediment Network, SedNet (www.sednet.org), of which this book represents one, focusing specifically at some of the key requirements and considerations needed for sediment management at the basin scale. The other books in the series consider sediment quality and impact assessment [57], sediment and dredged material treatment [58], and sediment risk management and communication [2], and the reader is strongly encouraged to consult these books to obtain detailed information on specific aspects of sediment assessment and management.

This book considers some of the main types of information needed to manage sediment at the river basin scale. Due to the complexity and often large size of river basins – the Danube basin, for example, is ca. 800,000 km^2 and composed of 19 countries – it is clearly not possible to cover all considerations and requirements for all river basin types in the limited space of this book. However, some generic principles are likely to be relevant to most basins. Some key requirements in the decision-making process for sediment management at the basin scale include:

- Identifying the drivers for sediment management. In other words, why does sediment need to be managed? There are a variety of drivers and pressures that operate at different spatial scales (Figure 11, also see [1, 59]). In most situations, sediment management is influenced and guided by legislation and policy, and this is discussed in Chapter 3 (this book). At the river basin scale, it is likely that there are many types of legislation and policy relevant for soils, water and waste where sediment plays a key, if often unstated, role. There are also non-regulatory drivers, such as agri-environment schemes, which influence how and why sediment is managed at a local and regional level. While local, site-specific management actions do not necessarily require an understanding and appreciation of all types of legislation, at the river basin scale they become relevant and need to be assessed in order to identify those that are relevant.
- Identifying the sources, pathways and transport processes of sediment and contaminants within the basin of interest. This is a prime need for sustainable and effective sediment management, by providing an understanding of how the sediment–contaminant system is behaving, and is a central requirement for source control as a management option. Chapter 4 (this book) provides a review of most of the main sources, pathways and transport processes for sediment in river basins, and includes both rural and urban systems. What is clear is that, at the basin scale, there are multiple sources of both sediment and contaminants, and that these sources supply sediment and/or contaminants at different parts of the basin and over different timescales.

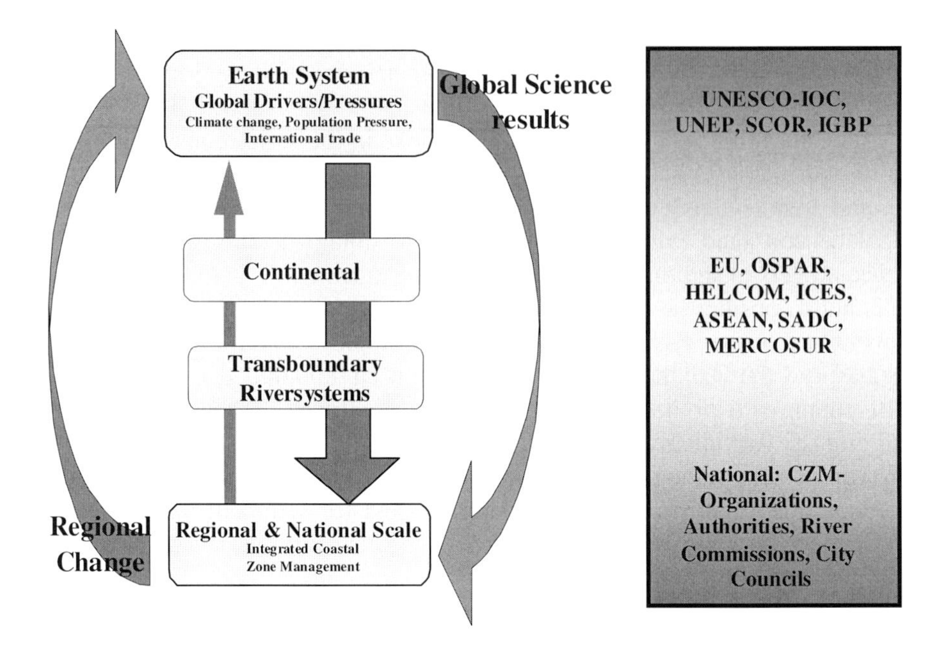

Figure 11. Different spatial scales of drivers and pressures relating to sediment management. See Chapter 3 (this book) for explanation of many of the initiatives listed in the right hand part of the diagram (source: [40], reproduced with permission of the author)

Similarly, pathways and transport processes are active at different times, with some being active essentially continually and others being 'switched-on' by specific triggering mechanisms such as rainfall events. In addition, there are various natural and anthropogenic activities which are modifying these sources, pathways and transport processes which need to be assessed.

- Using appropriate tools to assemble the relevant information and data needed for informed decision-making by managers. In many respects, the selection of which 'tool' to use in order to obtain the necessary information is dependent on the management question being asked, such as: What are the main sediment and contaminant sources? Where are they located in the basin? How will a particular management option (i.e. dredging) affect future sediment fluxes in the basin? Chapter 5 (this book) reviews many of the tools and techniques (such as monitoring, modelling and tracing techniques) that are available. Such tools provide much of the basic information that is required by many of the other aspects of the decision-making process that are discussed in the book. Thus, for example, tracing techniques provide information on sediment sources and pathways, while system modelling is often used to inform policy development through scenario analysis. Specific tools and approaches available to help river basin managers with decision-making that are particularly relevant at the river basin scale are risk assessment and cost-benefit analysis. The former is addressed within a separate book by Heise [2], and is also discussed in Chapter 2 (this book). The latter, and more specifically societal cost-benefit analysis, is described in Chapter 6 (this book). These are specific tools that can be used to assess and evaluate the various management options available to managers.

- Involving stakeholders in the decision-making process, from start to finish. This is now recognized as an important part of environmental management where there are various interested parties, often with conflicting interests and goals, and when there are several management options available. Indeed, stakeholder participation, and appropriate communication, is becoming increasingly incorporated within environmental legislation. Stakeholder participation, as part of the sediment management process, is discussed in Chapter 7 (this book) (see also [60–62]).
- Because of the complexity of trying to manage sediment at a large scale, such as a river basin, it is often useful to develop a framework (or nested frameworks) which incorporates many of the requirements and considerations listed above, as well as other important issues (see [2, 57, 58]). Examples of the types of frameworks that exist specifically for sediment, and also that could be adapted for sediment, are described in Chapter 2 (this book), and are also considered in Chapter 8 (this book). The latter also provides some suggestions for future requirements for sediment management at the river basin scale.

The following chapters cover the issues and requirements listed above in the context of sediment management at the river basin scale.

Acknowledgements

I would like to thank Mike Demissie (Illinois State Water Survey, USA), Ian Droppo (Environment Canada), Artyom Gusarov (Kazan State University, Russia), Ellen Petticrew (University of Northern British Columbia, UNBC, Canada) and Chris Thompson (University of New South Wales-ADFA, Australia) for supplying diagrams and allowing permission to reproduce them (Figures 10, 2, 6, 1 and 7/9, respectively), and Mike Demissie, Art Horowitz (USGS, USA) and Bill Renwick (Miami University, USA) for providing information and documents. Thanks are also extended to Ellen Petticrew and Neil Williams (both UNBC, Canada) for comments on an earlier version of this chapter.

References

1. Salomons W, Brils J (Eds) (2004). Contaminated Sediment in European River Basins. European Sediment Research Network, SedNet. TNO Den Helder, The Netherlands.
2. Heise S (Ed.) (2007). Sustainable Management of Sediment Resources: Sediment Risk Management and Communication. Elsevier, Amsterdam.
3. Droppo IG, Nackaerts K, Walling DE, Williams N (2005). Can flocs and water stable soil aggregates be differentiated within fluvial systems? Catena, 60: 1–18.
4. Droppo IG, Stone M (1994). In-channel surficial fine-grained sediment laminae (Part 1): physical characteristics and formation processes. Hydrological Processes, 8: 101–111.
5. Droppo IG (2001). Rethinking what constitutes suspended sediment. Hydrological Processes, 15: 1551–1564.
6. Petticrew EL, Arocena JM (2003). Organic matter composition of gravel stored sediments from salmon bearing streams. Hydrobiologia, 494: 17–24.
7. Petticrew EL (2006). The physical and biological influence of spawning fish on fine sediment transport and storage. In: Owens PN, Collins AJ (Eds), Soil Erosion and Sediment Redistribution in River Catchments: Measurement, Modelling and Management. CAB International, Wallingford, pp. 112–127.
8. Droppo IG, Leppard GG (2004). Sediment-contaminant interactions and transport: a new perspective. In: Golosov V, Belyaev V, Walling DE (Eds), Sediment Transfer through the Fluvial System. IAHS Publication 288. IAHS Press, Wallingford, pp. 429–436.
9. Gordon ND, McMahon TA, Finlayson BL (1992). Stream Hydrology: An Introduction for Ecologists. Wiley, London.
10. Townsend I, Whitehead P (2003). A preliminary net sediment budget for the Humber estuary. The Science of the Total Environment, 314–316: 755–767.
11. Hejduk L, Hejduk A, Banasik K (2006). Suspended sediment transport during rainfall and snowmelt-rainfall floods in a small lowland catchment, Central Poland. In: Owens PN, Collins AJ (Eds), Soil Erosion and Sediment Redistribution in River Catchments: Measurement, Modelling and Management. CAB International, Wallingford, pp. 94–100.
12. Bergmann H, Maass V (2007). Sediment regulations and monitoring programmes in Europe. In: Heise S (Ed.), Sustainable Management of Sediment Resources: Sediment Risk Management and Communication. Elsevier, Amsterdam, pp. 207–231.
13. Parker A, Bergmann H, Heininger P, Leeks GJL, Old GH (2007). Sampling of sediment and suspended matter. In: Barceló D, Petrovic M (Eds), Sustainable Management of Sediment Resources: Sediment Quality and Impact Assessment of Pollutants. Elsevier, Amsterdam, pp. 1–34.
14. Walling DE (2006). Human impact on land–ocean sediment transfer by the world's rivers. Geomorphology, 79: 192–216.
15. Slaymaker O (2003). The sediment budget as a conceptual framework and management tool. Hydrobiologia, 494: 71–82.
16. Slaymaker O (2004). Mass balance of sediment, solutes and nutrients: the need for greater integration. Journal of Coastal Research, SI(43): 109–123.
17. Owens PN, Slaymaker O (Eds) (2004). Mountain Geomorphology. Arnold, London.
18. Gomez B (1995). Bedload transport and changing grain size. In: Gurnell A, Petts G (Eds), Changing River Channels. John Wiley, Chichester, pp. 177–199.
19. Horowitz AJ, Kent AE, Smith JJ (in press). Monitoring urban impacts on suspended sediment, trace element, and nutrient fluxes within the City of Atlanta, Georgia, USA: program design, methodological considerations and initial results. Hydrological Processes.
20. Syvitski JPM, Peckham SD, Hilberman R, Mulder T (2003). Predicting terrestrial flux of sediment to the global ocean: a planetary perspective. Sedimentary Geology, 162: 5–24.

21. Trimble SW (1981). Changes in sediment storage in the Coon Creek Basin, Driftless Area, Wisconsin, 1853–1975. Science, 214: 181–183.
22. Owens PN (2005). Conceptual models and budgets for sediment management at the river basin scale. Journal of Soils and Sediments, 5: 201–212.
23. Foster IDL (2006). Lakes and reservoirs in the sediment delivery system: reconstructing sediment yields. In: Owens PN, Collins AJ (Eds), Soil Erosion and Sediment Redistribution in River Catchments: Measurement, Modelling and Management. CAB International, Wallingford, pp. 128–142.
24. White WR (2001). Evacuation of Sediments from Reservoirs. Thomas Telford Publishing, London.
25. Vörösmarty CJ, Meybeck M, Fekete B, Sharma K, Green P, Syvitski JPM (2003). Anthropogenic sediment retention: major global impact from registered reservoir impoundments. Global and Planetary Change, 39: 169–190.
26. Syvitski JPM, Vörösmarty CJ, Kettner AJ, Green P (2005). Impact of humans on the flux of terrestrial sediment to the global coastal ocean. Science, 308: 376–380.
27. Owens PN (2005). Soil erosion and sediment fluxes in river basins: the influence of anthropogenic activities and climate change. In: Lens P, Grotenhuis T, Malina G, Tabak H (Eds), Soil and Sediment Remediation. IWA Publishing, London, pp. 418–433.
28. Glymph M (1954). Studies of sediment yield from watersheds. IAHS Publication 36. IAHS Press, Wallingford, pp. 173–191.
29. Walling DE (1983). The sediment delivery problem. Journal of Hydrology, 69: 209–237.
30. Parsons AJ, Wainwright J, Brazier RE, Powell DM (2006). Is sediment delivery a fallacy? Earth Surface Processes and Landforms, 31: 1325–1328.
31. Panin A (2004). Land–ocean sediment transfer in palaeotimes, and implications for present day natural fluvial fluxes. In: Golosov V, Belyaev V, Walling DE (Eds), Sediment Transfer through the Fluvial System. IAHS Publication 288. IAHS Press, Wallingford, pp. 115–124.
32. Dedkov AP, Gusarov AV (2006). Suspended sediment yield from continents into the World Ocean: spatial and temporal changeability. In: Rowan JS, Duck RW, Werrity A (Eds), Sediment Dynamics and the Hydromorphology of Fluvial Systems. IAHS Publication 306. IAHS Press, Wallingford, pp. 3–11.
33. Caine N (2004). Mechanical and chemical denudation in mountain systems. In: Owens PN, Slaymaker O (Eds), Mountain Geomorphology. Arnold, London, pp. 132–152.
34. Dearing JA, Jones RT (2003). Coupling temporal and spatial dimensions of global sediment flux through lake and marine sediment records. Global and Planetary Change, 39: 147–168.
35. Milliman JD (1990). Fluvial sediment in coastal seas: flux and fate. Nature and Resources, 26: 12–22.
36. FAO (2006). FAO/AGL World river sediment yields database. http://fao.org/ag/agl/aglw/sediment/default.asp (accessed January 2006).
37. EUROSION (2004). Living with coastal erosion in Europe: sediment and space for sustainability. Part II – Maps and statistics. http://www.eurosion.org/reports-online/part2.pdf (accessed January 2006).
38. Owens PN, Batalla RJ (2003). A first attempt to approximate Europe's sediment budget. SedNet Report (www.sednet.org).
39. Owens PN, Batalla RJ, Collins AJ, Gomez B, Hicks DM, Horowitz AJ, Kondolf GM, Marden M, Page MJ, Peacock DH, Petticrew EL, Salomons W, Trustrum NA (2005). Fine-grained sediment in river systems: environmental significance and management issues. River Research and Applications, 21: 693–717.
40. Salomons W (2004). European catchments: catchment changes and their impact on the coast. EUROCAT Report. Vrije University, Amsterdam.

41. Burcher CL, Benfield EF (2006). Physical and biological responses of streams to suburbanization of historically agricultural watersheds. Journal of the North American Benthological Society, 25: 356–369.
42. Cotton JA, Wharton G, Bass JAB, Heppell CM, Wotton RS (2006). The effects of seasonal changes to in-stream vegetation cover on patterns of flow and accumulation of sediment. Geomorphology, 77: 320–334.
43. Thompson CJ (2006). The geomorphology of south east Australian mountain streams. PhD thesis. ADFA-UNSW, University of New South Wales, Canberra.
44. Newcombe CP, MacDonald DD (1991). Effects of suspended sediments on aquatic ecosystems. North American Journal of Fisheries Management, 11: 72–82.
45. Ryan PA (1991). Environmental effects of sediment on New Zealand streams: a review. New Zealand Journal of Marine and Freshwater Research, 25: 207–221.
46. Newcombe CP, Jensen JOT (1996). Channel suspended sediment and fisheries: a synthesis for quantitative assessment of risk and impact. North American Journal of Fisheries Management, 16: 693–727.
47. Armstrong JD, Kemp PS, Kennedy GJA, Ladle M, Milner NJ (2003). Habitat requirements of Atlantic salmon and brown trout in rivers and streams. Fisheries Research, 62: 143–170.
48. Petticrew EL, Rex JF (2006). The importance of temporal changes in gravel-stored fine sediment on habitat conditions in a salmon spawning stream. In: Rowan JS, Duck RW, Werrity A (Eds), Sediment Dynamics and the Hydromorphology of Fluvial Systems. IAHS Publication 306. IAHS Press, Wallingford, pp. 434–441.
49. Batalla RJ (2003). Sediment deficit in rivers caused by dams and instream gravel mining. A review with examples from NE Spain. Revista Cuaternario y Geomorfologia, 17: 79–91.
50. Kondolf GM (1994). Geomorphic and environmental effects of instream gravel mining. Landscape and Urban Planning, 28: 225–243.
51. Owens PN, Apitz S, Batalla R, Collins A, Eisma M, Glindemann H, Hoonstra S, Köthe H, Quinton J, Taylor K, Westrich B, White S, Wilkinson H (2004). Sediment management at the river basin scale: synthesis of SedNet Working Group 2 outcomes. Journal of Soils and Sediments, 4: 219–222.
52. Biever C (2006). Why splitting the Mississippi won't save the wetlands. New Scientist, 2571: 10–11.
53. Turner RE, Baustian JJ, Swenson EM, Spicer JS (2006). Wetland sedimentation from Hurricanes Katrina and Rita. Science, DOI 10.1126/science.1129116.
54. Demissie M, Xia R, Keefer L, Bhowmik N (2004). A sediment budget of the Illinois River. Illinois State Water Survey Contract Report 2004–13, Champaign, IL.
55. Demissie M, Keefer L, Slowikowski J, Stevenson K (2006). Evaluating the effectiveness of the Illinois River Conservation Reserve Enhancement Program in reducing sediment delivery. In: Rowan JS, Duck RW, Werrity A (Eds), Sediment Dynamics and the Hydromorphology of Fluvial Systems. IAHS Publication 306. IAHS Press, Wallingford, pp. 295–303.
56. Förstner U, Owens PN (2007). Sediment quantity and quality issues in river basins. In: Westrich B, Förstner U (Eds), Sediment Dynamics and Pollutant Mobility in Rivers. Springer, Berlin, pp. 1–15.
57. Barceló D, Petrovic M (Eds) (2007). Sustainable Management of Sediment Resources: Sediment Quality and Impact Assessment of Pollutants. Elsevier, Amsterdam.
58. Bortone G, Palumbo L (Eds) (2007). Sustainable Management of Sediment Resources: Sediment and Dredged Material Treatment. Elsevier, Amsterdam.
59. Owens PN, Collins AJ (2006). Soil erosion and sediment redistribution in river catchments: summary, outlook and future requirements. In: Owens PN, Collins AJ (Eds), Soil Erosion and Sediment Redistribution in River Catchments: Measurement, Modelling and Management. CAB International, Wallingford, pp. 297–317.

60. Gerrits L, Edelenbos J (2004). Management of sediments through stakeholder involvement. Journal of Soils and Sediments, 4: 239–246.
61. Morgan RPC (2006). Managing sediment in the landscape: current practices and future vision. In: Owens PN, Collins AJ (Eds), Soil Erosion and Sediment Redistribution in River Catchments: Measurement, Modelling and Management. CAB International, Wallingford, pp. 287–293.
62. Ellen GJ, Gerrits L, Slob AFL (2007). Risk perception and communication. In: Heise S (Ed.), Sustainable Management of Sediment Resources: Sediment Risk Management and Communication. Elsevier, Amsterdam, pp. 233–247.

Edited by Philip N. Owens

Conceptual and Strategic Frameworks for Sediment Management at the River Basin Scale

Sue M. White[a] and Sabine E. Apitz[b]

[a]*Cranfield University, Cranfield, Bedfordshire MK43 0AL, UK*
[b]*SEA Environmental Decisions Ltd, Little Hadham, Hertfordshire SG11 2AT, UK*

1. Sediment management today

Historically, sediment has been managed at the local level, in locations where it causes a problem. The main driver has normally been a specific local issue such as the need to maintain navigation, water storage or conveyance capacity, or, less frequently, the need to move contaminated sediment from a particular area or to restore habitat. Alongside this there is a long history of beneficial use of sediment, such as extraction of gravel or sand for the construction industry. Over recent years, increased awareness of the quality, as well as the quantity, aspects of sediment have led to increasingly stringent controls on sediment-related activities, and in particular on disposal of dredged material. As well as the more traditional licensing of, or good practice recommendations for, sediment extraction, sediment management activities are now subject to a range of legislation and guidance, such as the EU Hazardous Waste Directive ([1], see also Chapter 3, this book). Some dredged material in Europe is so heavily contaminated by a mixture of organic and non-organic pollutants (see e.g. [2–4]) that as soon as the extracted sediment breaks the water surface it is classed as a hazardous waste. This can lead to a difficult and expensive series of disposal strategies, such as dewatering, treatment or disposal to specialist contaminated disposal sites [5]. Even where the sediment is not classified as hazardous, increasing demands on agriculture to reduce diffuse sources of pollution and increased restriction on disposal of materials in the marine environment are also curtailing the traditional disposal routes to land or sea [6, 7]. This is making dredging activity increasingly (sometimes prohibitively) expensive, and it is argued that those with responsibility for dredging are being required to pay for polluting activities upstream in the river basin, for which they were not responsible, and which may have happened decades ago [8, 9]. Ultimately, increased costs associated with dredging impinge on economic activity and growth [10].

Sediment supply and transfer are also fundamentally affected by a range of activities that are not explicitly recognized as being related to sediment. Examples are the construction, rehabilitation or dismantling of weirs, dams [11–13], barrages and sluices within river systems, agriculture, construction projects, mining, forest logging activities [14], flood protection schemes [15], reduced river flow (due to abstraction, for example) and many others. These activities can either increase or decrease overall sediment supply to a stretch of river, or can change the characteristics of supplied sediment or the ability of the river to transport it (see Chapter 1, this book). Within Europe there is almost no length of river that has not been affected by at least one of these activities, and thus the sediment regime of all rivers can be considered to be anthropogenically affected [16]. The Environment Agency of England and Wales has assessed all water bodies in terms of whether they are at risk of failing the 'good ecological status' required by the Water Framework Directive (WFD) for a range of potential pressures: morphological change, which by definition involves alteration to the river environment and hence its sediment transfer regime, is one of these. Almost 60% of rivers, over 50% of lakes, almost 90% of estuaries and over 90% of coastal waters are classified as being at risk or probably at risk [17].

Sediment management actions are normally taken as a result of sediment quantity imbalances; however, the issues that make such management complex are often related to sediment quality. A wide range of contaminants have low solubility but high sorption potential, meaning that they bind to sediment particles, and in particular to the finer grained clay and silt particles [18]. Some contaminants are themselves particulate in nature and form a part of the sediment load of rivers. These bound or particulate contaminants can come from a range of sources, including bedrock, industrial discharges, agricultural practices, transport infrastructure and vehicles, mining waste and wastewater treatment plants (see Chapter 4, this book). Because of the intermittent way in which sediments, and their associated contaminants, move through the river basin and river system, contaminated sediments can be delayed within the sediment supply and transfer chain for decades or even centuries, remaining within alluvial floodplains or buried in sediment deposits (see Chapter 1, this book). When these sediment stores are disturbed, through extreme flow events, channel migration, alterations to the flow duration curve or direct physical disturbance, contaminants, some of which may not be currently used or may be banned from use, can be remobilized. There is thus a legacy problem, where sediments are acting as the memory of previous polluting activities in the river basin.

2. Understanding process: the need for a conceptual model

Appreciation of sediment as part of a dynamic river basin system, and of sediment as a 'memory' of previous activities, leads to the conclusion that it may not be most effectively managed at an individual site. Sediment also needs to be more explicitly considered in a range of activities within river basins which may affect the river sediment regime whilst being targeted towards quite different ends. It is thus important to consider sediment management within its wider environmental, economic and social context. A good parallel here is the Shoreline Management Plans (SMPs) process [19, 20] developed for management of the UK coastline, where the coast is divided into a number of discrete sediment cells through which sediment moves but beyond which it does not. These are analogous to some extent with river basins. For the SMPs, activities within a sediment cell must be considered in terms of their impact on the rest of the cell. Thus stabilization of one part of the coast that cuts off sediment supply to other points and causes them to erode, or a development that activates sediment sources, should not be undertaken without full understanding of the implications and how they may also be managed. This has enabled an ambitious forward-looking study, 'Future coast' [21], which envisages changes in coastal dynamics under climate and social change scenarios.

In order for river basins to be used as sediment management units, it is vital to have a conceptual model of river basin functioning that links different areas in space and time, and allows potential consequences (impacts) of drivers to be evaluated. This can be visualized in various ways. For instance, Figure 1 gives examples of actions and reactions along an energy gradient (slope) where sediment is present. The shaded blocks may represent a continuum of soil and sediment from a hillside to a river, along a river stretch or from top to bottom of a river basin. The lighter shade represents sediment 'blocks' that may be a field, an alluvial plain, a hillslope, etc., and that are uncontaminated. The dark shaded 'blocks' represent sediment that is contaminated by a compound that is ecologically or environmentally undesirable. Thus we may think of an area of land with high heavy metal content, or a river bed deposit with high phosphate concentration. The quantity and quality labels indicate why we might take action and what sort of impact it would have. Thus for the middle example (quantity-quality), we may remove clean sediment because we need to increase flood conveyance (a quantity issue), but the consequence is exposed or mobilized contaminated sediment downstream (a quality issue).

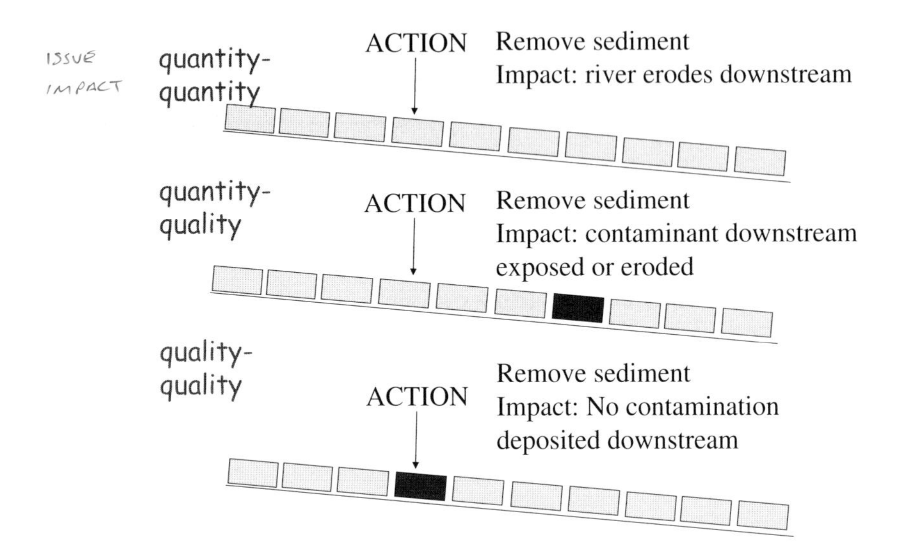

Figure 1. Illustration of the potential impacts of various types of sediment activities on sites further down an energy gradient (source: [22] reproduced with the permission of Ecomed Publishers)

So far the consequences of human actions in respect of Figure 1 have been discussed. However, the mobilization of sediment at the 'ACTION' point may be due to an extreme rainfall or flow event, or due to some indirectly related factor, such as a release of water from an upstream dam. Thus even where no direct sediment management activities are planned, it is important to understand how different soil and sediment bodies are connected in space.

Figure 2 presents similar information in terms of both sediment quantity and an evaluation of risk posed by sediment in a particular location based on its relative energy (topographic) position. The darker shades on the top layer of the diagram represent more contaminated soil or sediment, and the lighter shades represent cleaner sediment. The lower surface represents the potential energy of the soil or sediment at each point; at its simplest this could be thought of as topography. A relative evaluation of sediment at points A, B, C and D can be carried out. For example, A is a site with contaminated soil or sediment that is upstream (up the energy gradient) of the cleaner sediment site B, and of the contaminated site at C. On the other hand, D is a clean soil or sediment site, which is upstream of a more contaminated site, C. Traditionally, most sediment management activities have taken place at the lower end of river basins, where sediment is deposited. However, consideration of the possible risks in this hypothetical basin would suggest that site A should be the primary focus for

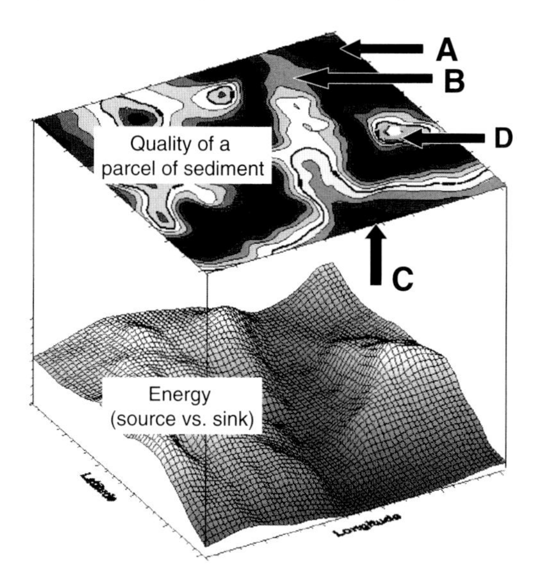

Figure 2. Conceptual diagram of a projection of sediment energy (source vs sink) and quality using data from a conceptual basin model to inform risk prioritization (source: [22] reproduced with the permission of Ecomed Publishers)

remediation, before site C. This is because A has the potential to both contaminate a clean site at B and to enhance the problems at C. Conversely, sediment movement from D to C could be a positive factor in that contaminants at C would be buried and could be subject to natural attenuation processes and may then not prove to be as great a risk in the future. Thus actions taken which reduce sediment supply from site D, such as soil conservation programmes, planting of hedgerows or buffer strips, may not prove to be beneficial at the basin level, in terms of contaminant risk. At least, consideration of contaminant impacts down the energy gradient should be considered.

An example of unforeseen consequences is the erosion of the river bed at sites along the River Danube ([23]; also see Chapter 8, this book). This has happened as a consequence of the construction of a series of hydroelectric barrages along the river, which trap sediment in the impounded water behind the barrage. As one moves downstream, the river becomes increasingly starved of sediment and begins to remobilize the river bed. In some places the erosion has been so severe as to directly connect the river with the underground aquifer, negating some of the benefits of groundwater due to long isolation from the land surface.

3. Sediment management at the river basin scale

As has been seen, sediment is not currently managed at the river basin scale, although this is exactly the scale over which sediment supply, transfer and deposition occurs (see Chapter 1, this book). The reasons for this lack of a basin-scale approach are many.

- There is a paucity of data on sediment flux in river systems. This is related to the highly event-based nature of sediment movement. Generally, for rivers in Europe, around 70–90% of sediment flux occurs in the top 20% of flows [24, 25]. There are few routine monitoring systems adequate to capture these events.
- The sediment problem or issue may occur at some considerable distance in space and time from the origin of the problem. The principle of 'polluter pays' is thus not one that can be readily adopted.
- There may be multiple sources of sediment and contaminant that have been active at various times and places. Thus the sediment 'problem' may be the result of myriad causal factors.
- Normally the sources of sediment and contaminant are not in areas that are owned or controlled by the problem owner. This means that problem owners feel powerless to resolve the situation.
- There has been little focus on sediment as part of a dynamic ecosystem at the river basin scale. In part, this has been because it is so difficult to identify and quantify all sources over space and time. Sediment-borne contaminants increase the system complexity further.
- There is little information on how much sediment, or what type of sediment, is 'good', i.e. sediment concentration or flux targets are largely undefined.

Sediment issues can be further subdivided into existing and identified, existing and unidentified, and potential future problems – all requiring different approaches and management strategies. To support basin-scale management, we must remember where in the decision process sediment issues fit in order to enhance communication and interaction.

For example, Apitz and Power [26] compare a framework designed for dredged material disposal, where:

- a management decision has already been made – to dredge;
- focus is on appropriate disposal methods or sites; and
- frameworks are highly prescriptive and often treaty-driven;

with a framework to determine the *in-situ* risk of contaminated sediments, where:

- it remains to be determined if risk exists;
- it remains to be determined if there is a need to manage at all; and
- frameworks are often complex, flexible and very site- or region-specific.

An existing identified need to dredge to maintain navigation requires site-specific risk assessment and, we argue, a consideration of the potential impacts on hydrologically connected areas. Here the approach may be as set out by Babut *et al.* [27].

An existing but unidentified problem may be an area of sediment deposition with associated contamination problems. Here a site-specific risk assessment may be carried out, again including possible impacts downstream if the sediment is remobilized. In reality, the best approach may be to do nothing more than monitor the situation, as contaminants may degrade over time or may be, or may become, biologically unavailable. A specific example of this, where a problem may be expected but not clearly identified, is in the case of historic mining activity. Macklin *et al.* [28] outline a geomorphological approach to the management of rivers contaminated by metal mining (also see Figure 6 in Chapter 8, this book). This includes a detailed review of mining activity, mapping of deposition areas, valley floor sediment sampling and analysis, comparison with acceptable contaminant thresholds, and analysis of likelihood of metal contamination to the river and finally action in key hazard areas. Economically, action in hazard areas, as opposed to identified risk areas, may be prohibitive and so an on-going programme of monitoring would be required.

In terms of future management then, an overall river basin approach is required (see Chapter 1, this book). This will predominantly involve ensuring that a correct sediment balance is achieved throughout the river system. This means that a range of spatially varying environmental benefits are targeted and achieved. This is the approach that needs to be incorporated into the River Basin Management Plans required by the WFD. Here we may use a broadly similar approach to that outlined by Macklin *et al.* [28] to identify potential problems or issues as presented in Apitz *et al.* [29] (Figure 3). An alternative conceptualization is presented in Figure 4. This would lead on to a number of site-specific assessments. At either of these stages we may decide that no further management is required. If action is required, then site-specific management would occur and the situation would be monitored to ensure that targets have been met. All of this would need to occur within a conceptual

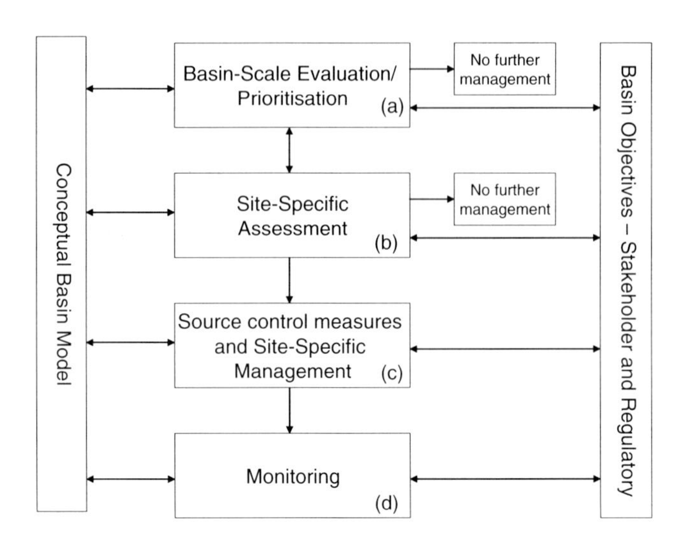

Figure 3. Conceptual diagram of the relationship between basin-scale and site-specific assessment and management in a river basin (source: [29] reproduced with permission of Elsevier)

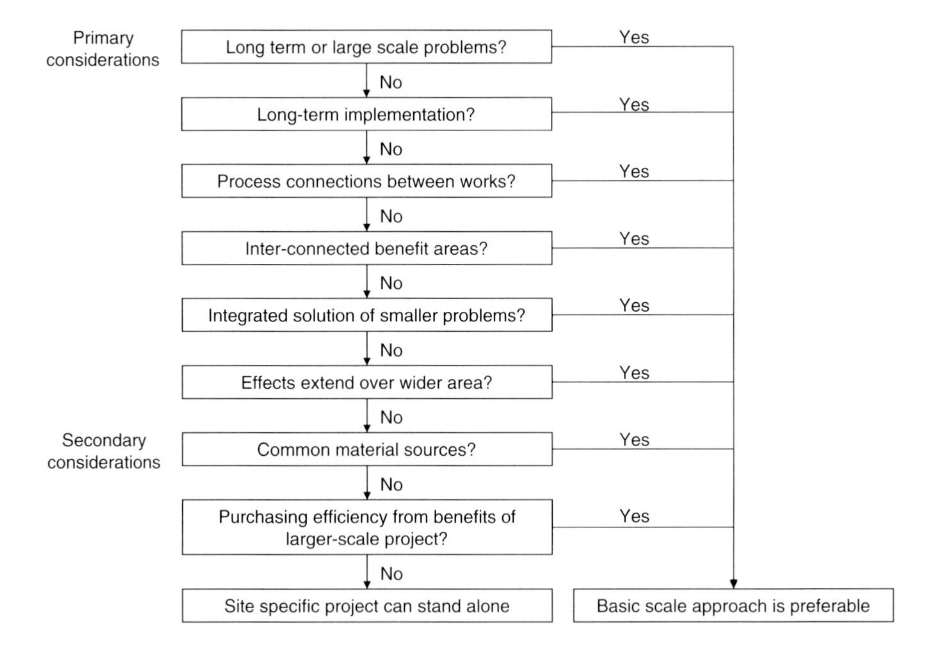

Figure 4. Decision tree for site-specific vs basin-scale management actions (source: modified from [32])

understanding of basin-scale sediment processes (see Chapter 1, this book; and [30]) and must relate to the regulatory and stakeholder objectives for the river basin (see Chapters 3 and 7, respectively, this book, and [31]). Such objectives may be economic, ecological or social, or a mixture of all of these.

Sediment management is also fragmented in other ways:

- by its relationship to sediment: researcher, problem owner, sediment producer, regulator, environmental protection organization (sediment management drivers);
- by discipline: toxicologist, chemist, geologist, engineer, ecologist, economist, politician;
- by environment: freshwater, estuarine, marine;
- by regulatory framework; and
- by geographical location (region, country).

It is, therefore, important that we develop management frameworks that can be viewed from, and by, all of these specializations and contexts. Part of this task involves development of a common vocabulary and tools for visualization and evaluation of sediment problems and potential solutions.

The dynamic nature of sediments and the international aspects of the problem call for a new approach (see Chapter 8, this book). This requires the involvement of many layers of political, technical, scientific, economic and environmental analysis, and also must involve very different drivers, organizations and approaches. It is, therefore, essential that we define strategic, conceptual and process frameworks that:

- identify these interactions;
- define common issues and terms; and
- expedite information exchange.

This needs to be done across organizations, across scales and across disciplines.

It is important that we remember that sediment is just one aspect of river basin management, where other aspects may be:

- water management (resource, flooding);
- biota/habitat management;
- agricultural production/policy;
- industry policy and control; and
- socio-economics.

Management of any of these aspects will have impacts on sediment dynamics within the basin. It is increasingly clear that we should not manage sediments, or anything else, alone – we should balance risks and goals in dynamic river basins.

4. Risk management in river basins

In practice, there are always limited resources available for management within river basins and, ideally, decisions about which issues will be addressed should be based on an assessment of risk. More detail on risk-based approaches for sediment management can be found in Heise [33].

4.1. Basin-scale risk framework

Long-term risk reduction needs a basin-scale decision framework. Because of the dynamic, complex and interconnected nature of sediments, from sources and rivers to estuaries and the sea (see Chapter 1, this book), effective and sustainable management strategies must focus on the entire sediment cycle, rather than on one unit of sediment at a time. Such an approach will help focus limited resources to maximize the achievement of management objectives, including basin-scale risk reduction. A basin-scale risk management framework should be comprised of two principal levels of decision-making; the first being a basin-scale evaluation (prioritization of sites for further evaluation and/or management) and the second being an assessment of specific sites for risks and management options (site-specific risk ranking and management), and this can be seen in differing detail in Figures 5 and 6.

Management activities at one site may significantly influence contaminant and sediment dynamics downstream and sometimes even upstream. Such effects can be extreme when rivers are dammed or canalized, or reservoirs are built, extensively changing suspended matter concentration, current velocity and thus erosion potential, flooding risk, etc. However, less dramatic activities such as the relocation of dredged material, deepening of waterways and restoration of floodplains can have impacts on sediment transport. Thus, due to the interconnectivity of sites and of the impacts of management activities, an extensive description of the sediment, contaminant and risk processes at the river basin scale in question is necessary to facilitate risk management. Such a description, called the Conceptual Basin Model (CBM), defines what is known or predicted about the mass flows of particles, e.g. suspended matter concentrations in the main river and its tributaries during normal and flood events, resuspension areas, settlement areas (and rates), etc. Information on predicted shear stresses and critical erosion thresholds, from which resuspension

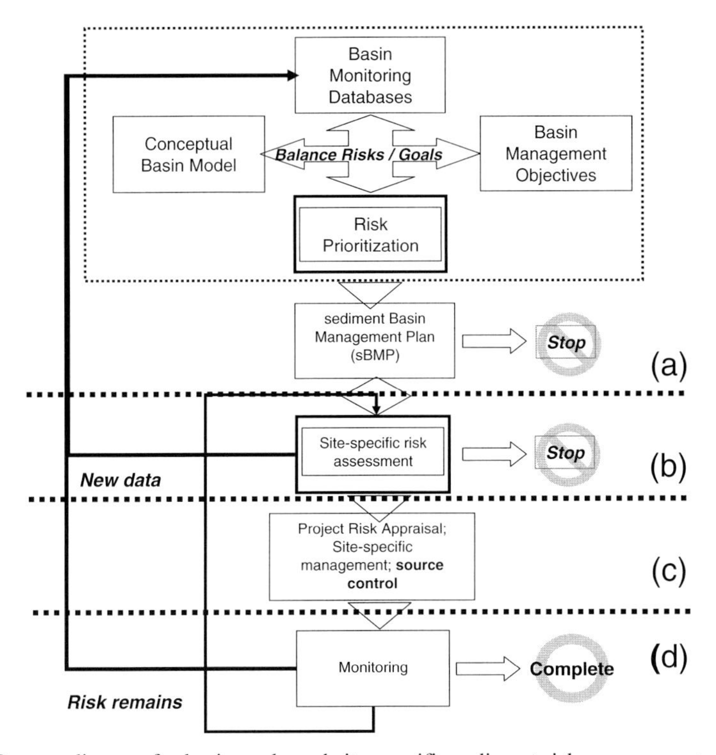

Figure 5. Process diagram for basin-scale and site-specific sediment risk management. A manager may 'enter' the process at the basin scale (in level 'a'), or at the site-specific scale (in level 'b'). Note: Whilst the focus of this figure is on sediment risk management, the same concepts could be applied to other media (source: [29] reproduced with the permission of Elsevier)

potentials could be deduced, is helpful in developing an informative CBM, but unfortunately scarce.

Sediment dynamics are strongly connected with risk, as they control the transport and exposure of particle-bound contaminants:

- river flow velocity and critical erosion thresholds determine resuspension of sediment and thus the potential remobilization of contaminants;
- reservoirs and other still water zones lead to removal of contaminated sediments from the dynamic system;
- concentration of suspended matter and its contaminant levels determine the contaminant load that is being transported downstream; and
- dilution of suspended matter at the confluence of tributaries can lead to a decrease in contaminant concentration below an effective threshold, or it can increase the volume of contaminated material due to mixing and add to the cocktail of toxic substances.

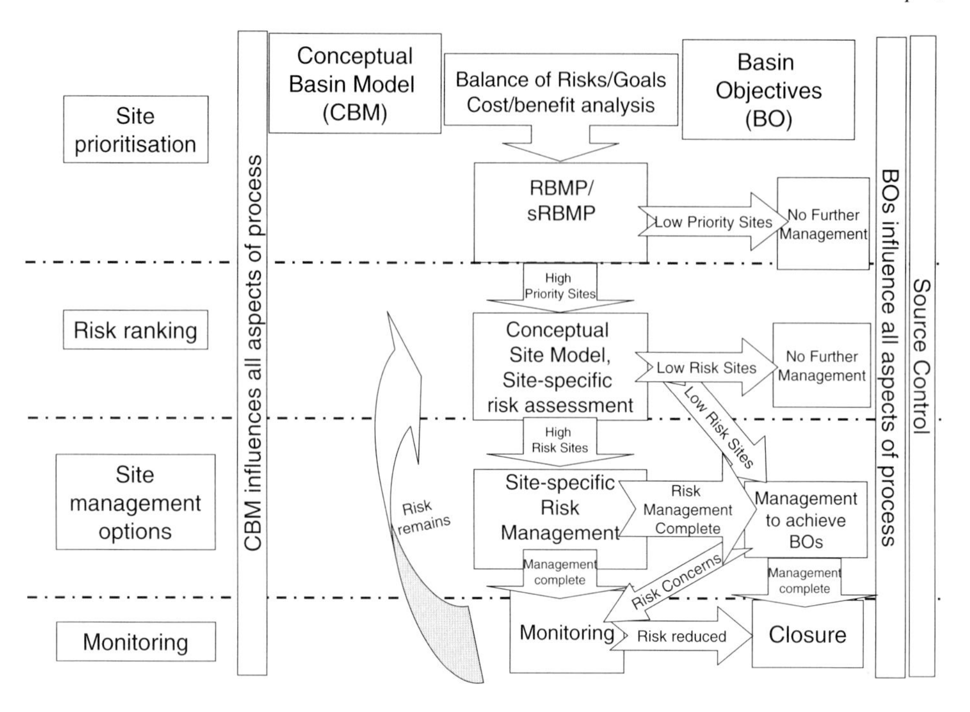

Figure 6. A risk-based sediment framework, identifying basin-scale and site-specific phases (source: [22] reproduced with the permission of Ecomed Publishers)

Thus, the CBM should also include information on the mass flow of contaminants, including an inventory of what substances are of concern in that river basin, their concentration in suspended matter and in sediment, and their potential sources of emission, whether they are point (industries, municipal WWTP) or diffuse sources (e.g. surface runoff, diffuse urban effects, atmospheric fall out), whether they are current (industrial emissions) or historic (secondary sources: historic contaminated sediment). For further information on sediment and contaminant sources see Chapter 4 (this book).

In practice we need to move through a series of steps to achieve successful 'closure' of a sediment issue. This involves two explicit stages of risk characterization:

- basin-scale site prioritization; and
- site-specific risk assessment.

4.2. Site prioritization at basin scale

If, by any process within the river basin, (contaminated) sediment poses a risk to one or more of the basin objectives, reduction of that risk becomes a necessary item in a sediment management action plan. In most cases, risk reduction is most effectively achieved if the whole river basin is addressed because of the dynamic, complex and interconnected nature of sediments. Various aspects of the CBM should provide screening-level information about how sediment and contaminant distribution and dynamics might possibly endanger the agreed-upon river basin objectives, as well as how these actions might be addressed. Data from the CBM, as well as a site risk prioritization and a consideration of the socio-economic and ecological basin objectives, when combined with other factors, will ultimately lead to a basin sediment management plan.

Site risk prioritization should be done according to a number of criteria, preferably quantifiable at a screening level, including: the site's *location along the up- and downstream gradient* (e.g. how are downstream sites affected by contaminated sediment sources upstream?), *quantity of contaminated sediment* (e.g. is the volume large enough to present a risk?), *evaluation of sediment quality* (e.g. how contaminated is the sediment?), *sediment dynamics* (e.g. is (contaminated) sediment transported downstream?) and the *potential expected risks or benefits* of proposed management actions at a given site and for the river basin (e.g. to what extent can the risk to the site and the basin be reduced when the site is managed?). In those cases in which sediment management decisions begin at the site-specific level, it is essential that the impacts of these actions upon adjacent sites and basin-scale quality and objectives are considered.

4.3. Site-specific risk ranking and management

If a site is identified as high priority during site prioritization, then it will be subject to a management process that includes site-specific risk ranking. A site-specific risk ranking is needed in order to determine, in greater detail, what the risks at a given site are. A tiered assessment is recommended that comprises, at different levels, the use of chemical, ecotoxicological and sediment community data in order to assess the *in-situ* risks and predict those that are connected with management activities. Such an approach requires the development of explicit measures of exposure, related to ecological processes, which must be selected based upon site-specific conditions and management options (or scenarios). Ultimately, basin-scale risk management will require the harmonization of risk assessment and ranking approaches. The selection of risk management or disposal approaches requires a comparative risk assessment that identifies (and possibly compares) the risks to the environment due to management options, such as dredging. Post-remedial monitoring, to confirm risk reduction, to flag continuing problems, and to update and refine CBMs is recommended as well.

5. Frameworks for sediment management in Europe

Within Europe, at the national and regional level, EU policy and legislation are now the main drivers for environmental management. Although sediment does not have dedicated legislation, it does provide an interface in a physical sense between many other legislative areas and particularly those EU directives that relate to waste/contaminants, water and soil, such as the Hazardous Waste, Soil, Habitats, Fisheries, Bathing Waters and Water Framework Directives (see Chapter 3, this book), as well as the REACH initiative for manufactured chemicals. With a change in the focus of water policy towards the river basin scale, mainly through the introduction of the WFD, there is now a need to recognize and consider not only the WFD (as a driver for management) but also a much broader range of policy and legislative drivers that relate to the different parts and functioning of a river basin (land, rivers, coasts, etc., also see below).

The identification and evaluation of the drivers for sediment processes and management at different scales, from local to river basin to EU, and the recognition of the importance and role played by drivers at the global scale, such as global market forces and global climate change, are an important requirement for more effective sediment management. An effective means of linking these drivers to ultimate impacts is the Driving force – Pressure – State – Impact – Response framework (DPSIR) used by the European Commission (EC), the Organization for Economic Cooperation and Development (OECD), the European Environment Agency (EEA) and others, to provide an overall mechanism for analysing environmental problems. This approach defines the interactions between these various parameters as:

- *driving forces*, such as industry, agriculture or climate, produce
- *pressures on the environment*, such as polluting emissions or more extreme events, which then degrade the
- *state of the environment*, which then
- *impacts on human and eco-system health*, causing society to
- *respond with various policy measures*, such as regulations, information and taxes, which can be directed at any other part of the system [34].

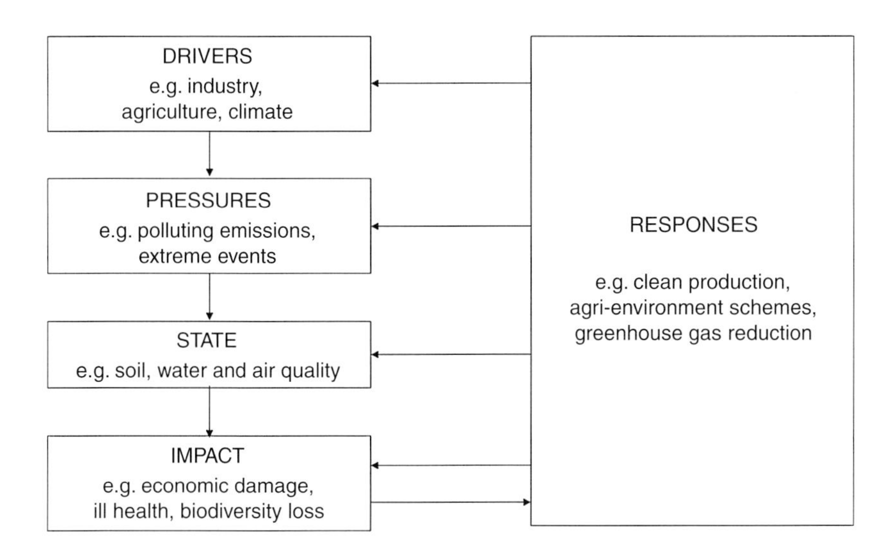

Figure 7. The DPSIR approach (adapted from [34])

Particularly useful for policy-makers, DPSIR offers a basis for analysing the inter-related factors that impact on the environment and the layers of policy targets for which we manage river basins. This conceptual approach can be illustrated as in Figure 7.

It is important to note that the DPSIR approach is just a tool that allows one to describe how processes affect each other, and how they inform decisions. While the figure looks simple, showing rather intuitive links between processes, the complexity lies in how the boxes link. To make the link between the parameters we can measure (States) and the risks we are concerned about (Impacts), or between acknowledged Impacts and the selection of appropriate Responses, often requires a complex set of models (also see Chapter 5, this book). For example, in contaminated sediment management, a direct comparison between sediment contaminant levels and target values may be used to infer toxicity, or a much more extensive, site-specific ecological risk assessment, using tiered approaches, triads or weight of evidence approaches may be applied [4]. Then, the selection of appropriate remedial responses may involve a complex comparative risk assessment, considering the financial, regulatory, scientific and technical aspects of the site. Appropriate approaches may be affected by the scale of the problem, the possibilities for source control and/or natural attenuation, costs, and many other factors. The success or failure of a response must also be evaluated over time to determine changes in drivers and pressures.

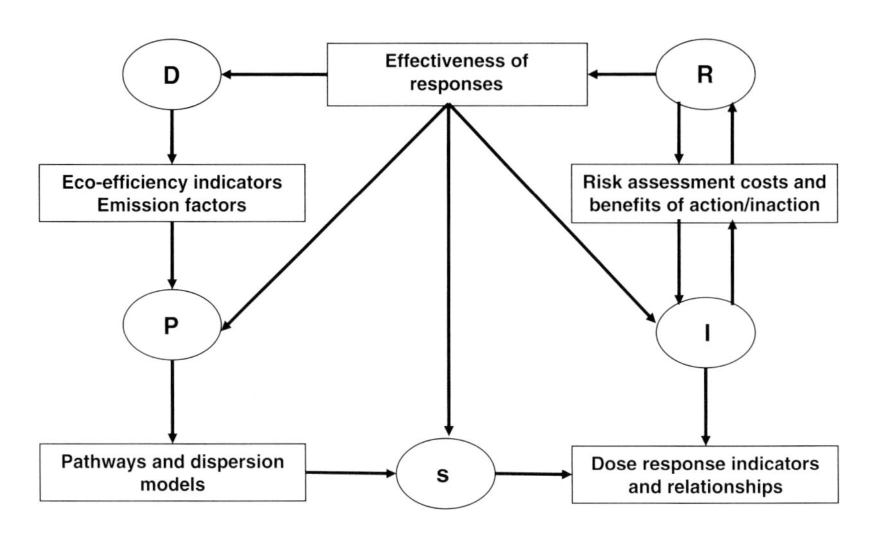

Figure 8. Indicators and information linking DPSIR elements (source: [35])

Without an explicit statement of how data and models are going to be used to inform decisions, possibly unnecessary and expensive studies and actions may result. To communicate these links to policy-makers, indicators of risk are used. According to Smeets and Weterings [35], 'Communication is the main function of indicators: they should enable or promote information exchange regarding the issue they address. Our body temperature is an example of an indicator we regularly use. It provides critical information on our physical condition. Likewise, environmental indicators provide information about phenomena that are regarded typical for and/or critical to environmental quality … that enable communication on environmental issues.'

Clearly, embedded in a well-designed and meaningful indicator is the conceptual model used to link processes within a DPSIR framework, but different sorts of indicators are used to link different boxes. Smeets and Weterings [35] provide an illustration to show what types of indicators (and, by implication, what types of tools and models) are needed to establish meaningful links in the conceptual process (Figure 8).

6. An example of a basin-wide sediment management strategy: the Norfolk Broads, England

The Norfolk Broads are a series of shallow lakes connected by rivers in eastern England (Figure 9). The lakes are highly valued for their macrophyte communities and as a habitat for a range of birds. They are thus classified and

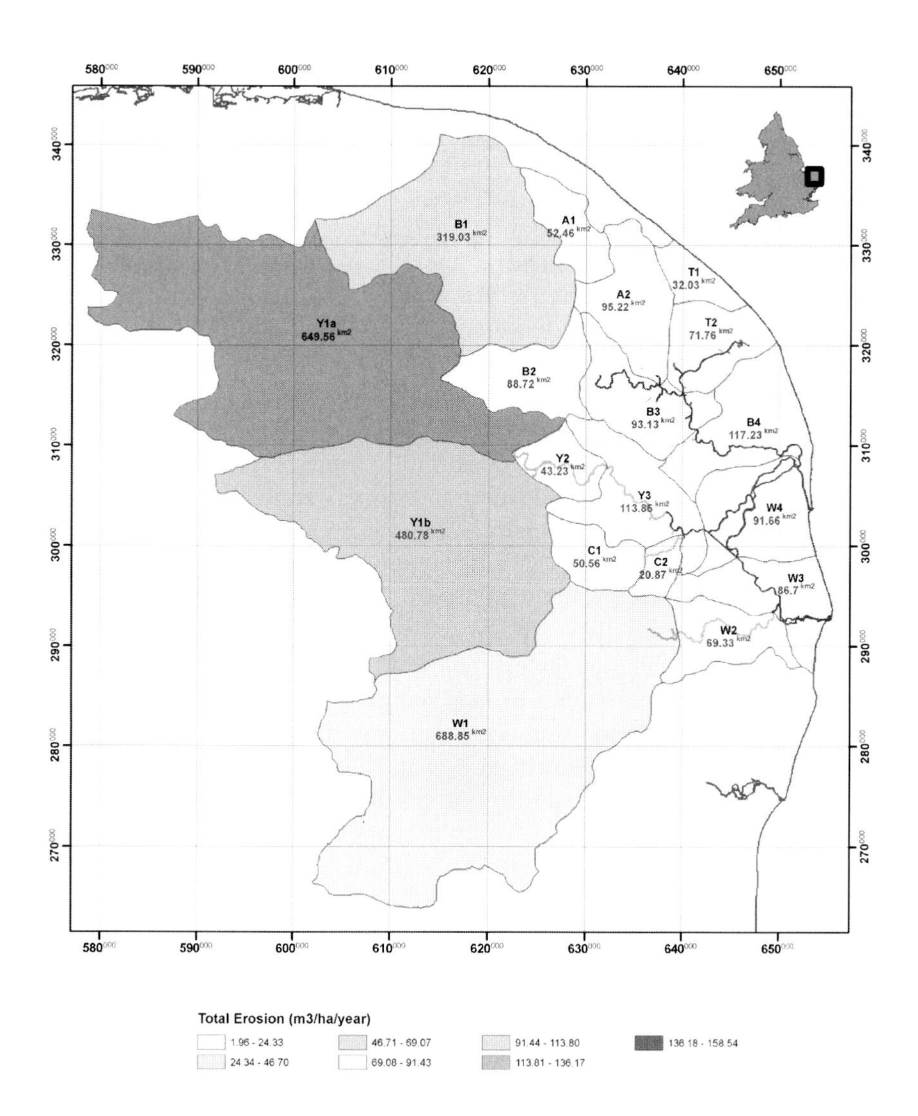

Figure 9. Broads river basins and management areas – all areas numbered 1, 1a, 1b lie outside the BA management area (source: [36])

protected under a range of environmental initiatives: RAMSAR; UK Sites of Special Scientific Interest (SSSI); and Habitats Directive Special Areas of Conservation (SAC). However, the lakes are situated in the middle of the most productive arable farming area in the UK, and this places particular demands on lake managers. The Norfolk Broads are managed by a statutory authority, the Broads Authority (BA), which has statutory duties relating to conservation, land and water management, planning, recreation and visitor services. The spatial extent of the BA management area is not contiguous with the hydrologically defined river basins, and thus the BA cannot directly control activities in a series of 'headwater basins' that drain into the Broads (Figure 9). Furthermore, they do not own the land within the Broads management area, and therefore have to enter into a number of voluntary agreements with local land managers.

One of the largest budget items for the BA is the dredging of sediment to maintain navigation, as the Broads are a popular area for leisure boating and sailing. In addition, some remedial dredging has taken place in one of the lakes in an attempt to remove sediment-bound phosphorus from the lake ecosystem. Many of the lakes are now not in the clear water, macrophyte-rich condition from which their ecological importance derives. In large part this is due to increasing nutrient levels, although salinity increases are also important. A decision was therefore taken to develop a sediment strategy for the Broads, the first phase of which was to attempt to quantify sediment sources [36].

The first step was to carry out a literature review and a consideration of possible sediment sources. This resulted in a conceptual model of the Broads sediment sources (Figure 10). An extensive exercise was then carried out to identify data sources and to make estimates of sediment inputs and outputs. This process was facilitated through a Broads sediment DPSIR analysis (Figure 11), which allowed interactions between various drivers and impacts to be investigated. All available information was then collated to produce a first sediment budget for the Broads (Figure 12). This is incomplete because for some expected sediment inputs very little quantitative evidence was available. The sediment budget plus the DPSIR analysis were then used to identify priority actions for sediment management for the BA. These ranged from priority areas of research and/or monitoring to allow quantification of those inputs and outputs that could not be adequately quantified [37], to influencing policy and activities in areas outside of direct BA control. This information has provided a basis for a sediment management strategy defined by the BA to prioritize areas of dredging, to allocate budgets effectively, and to obtain best possible outcomes for the wide range of stakeholder groups in the area [38, 39].

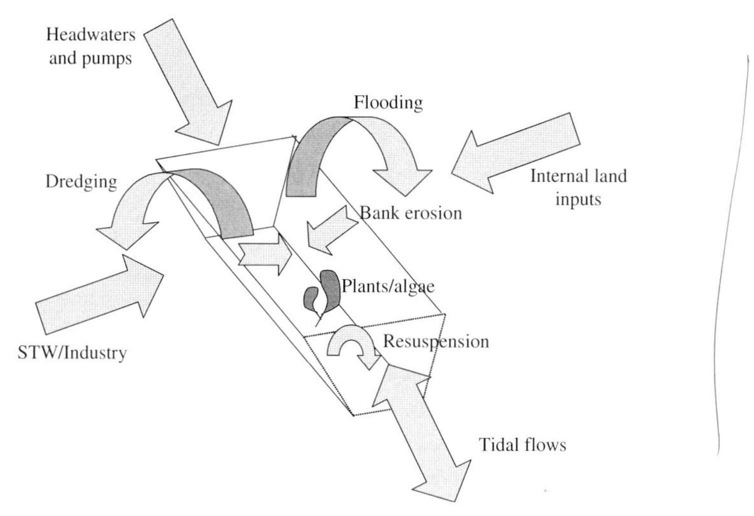

Figure 10. Conceptual model of potential sediment sources to the Broads: STW = sewage treatment works (source: [36])

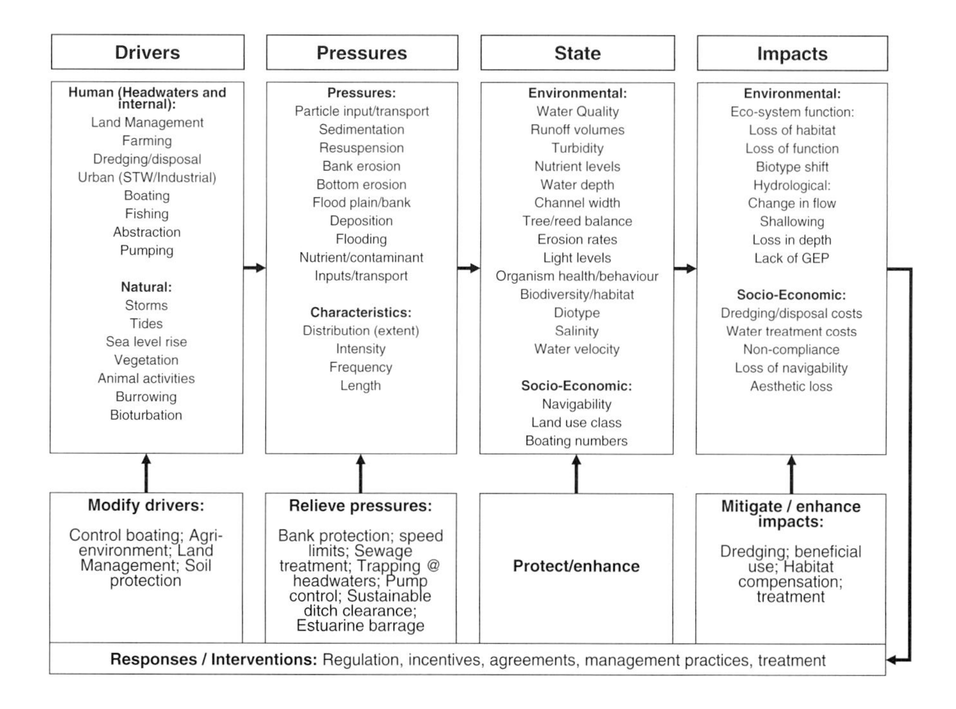

Figure 11. Sediment DPSIR for the Broads (source: [36])

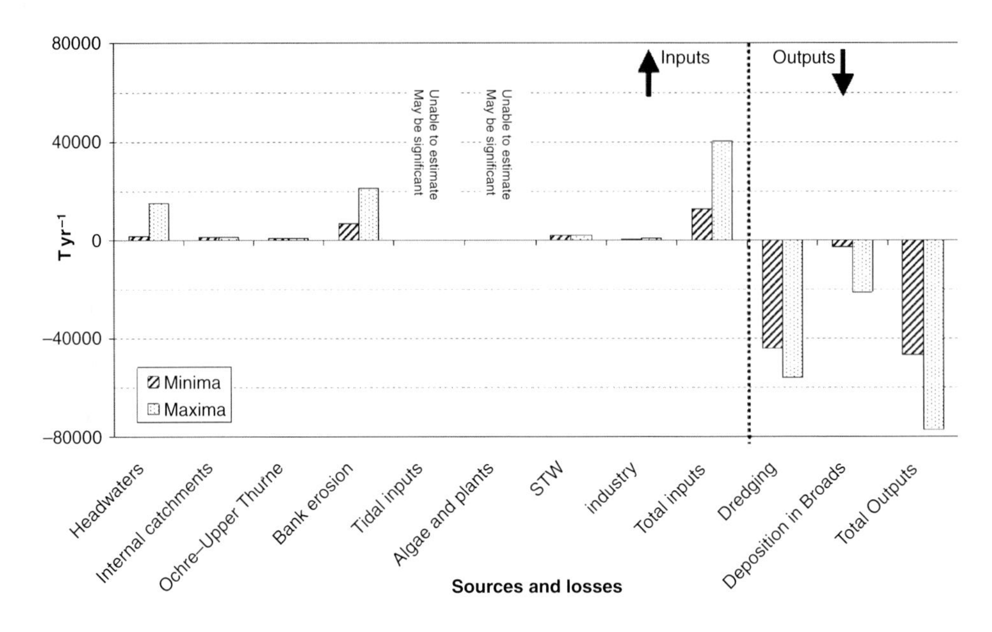

Figure 12. Sediment budget for the Broads management area (source: [36])

7. Conclusions

Ultimately, sediment management is about balancing risks and goals. This needs to happen in the context of the broader basin objectives and our understanding of basin functioning (the conceptual basin model). In particular, it must be remembered that:

- decisions will be informed by available data;
- the process is iterative, with post-intervention monitoring informing both the high-level basin understanding and the site-specific assessment of a problem; and
- the various stages in risk assessment and prioritization have been addressed within SedNet [33].

Finally, it must be remembered that sediment management is just one part of an overall basin management strategy. Sediment issues cut across many of the policy areas at European, national and basin level. Because of this, sediment needs to be considered as a part of many policy arenas. Ideally, site prioritization should be used to derive a ranked list of sediment problems within a basin. Site-specific sediment problems must be viewed within the basin context and as part of a multivariate basin management policy. Sediment management frameworks help us to understand the interactions, intersections and information exchanges necessary to manage sediment sustainably into the future. Many of these issues are considered further in Chapter 8 (this book), while the following chapter (Chapter 3) specifically reviews current and forthcoming legislative and policy frameworks for sediment management.

References

1. Apitz SE, Brils J, Marcomini A, Critto A, Agostini P, Michiletti C, Pippa R, Scanferla P, Zuin S, Lánczos T, Dercová K, Kočan A, Petrík J, Hucko P, Kusnir P (2006). Approaches and frameworks for managing contaminated sediments – a European perspective. In: Reible DD, Lanczos T (Eds), Assessment and Remediation of Contaminated Sediments. NATO Science Series IV, Earth and Environmental Sciences Vol. 73. Springer, Dordrecht, pp. 5–82.
2. Canellas-Bolta S, Strand R, Killie B (2005). Management of environmental uncertainty in maintenance dredging of polluted harbours in Norway. Water Science and Technology, 52: 93–98.
3. Cappuyns V, Swennen R, Devivier A (2006). Dredged river sediments: potential chemical time bombs? A case study. Water, Air and Soil Pollution, 171: 49–66.
4. Barceló D, Petrovic M (Eds) (2007). Sustainable Management of Sediment Resources: Sediment Quality and Impact Assessment of Pollutants. Elsevier, Amsterdam.
5. Bortone G, Palumbo L (Eds) (2007). Sustainable Management of Sediment Resources: Sediment and Dredged Material Treatment. Elsevier, Amsterdam.
6. Sigua GC, Holtkamp ML, Coleman SW (2004). Assessing the efficacy of dredged materials from Lake Panasoffkee, Florida: implication to environment and agriculture – Part 2: Pasture establishment and forage productivity. Environmental Science and Pollution Research, 11: 394–399.
7. Peijnenburg W, de Groot A, Jager T, Posthma L (2005). Short-term ecological risks of depositing contaminated sediment on arable soil. Ecotoxicology and Environmental Safety, 60: 1–14.
8. Hansen LT, Breneman VE, Davison CW, Dicken CW (2002). The cost of soil erosion to downstream navigation. Journal of Soil and Water Conservation, 57: 205–212.
9. Schafer J, Blanc G, Audry S, Cossa D, Bossy C (2006). Mercury in the Lot-Garonne River system (France): sources, fluxes and anthropogenic component. Applied Geochemistry, 21: 515–527.
10. Grigalunas TA, Opaluch JJ, Chang YT (2005). Who gains and who pays for channel deepening? The proposed Delaware channel project in Delaware. Inland Waterways, Ports and Channels and the Marine Environment. Transportation Research Record, 1909: pp. 62–69.
11. Davidson GR, Bennett SJ, Beard WC, Waldo P (2005). Trace elements in sediments of an aging reservoir in rural Mississippi: potential for mobilization following dredging. Water, Air and Soil Pollution, 163: 281–292.
12. Cui YT, Parker G, Braudrick C, Dietrich WE, Cluer B (2006a). Dam Removal Express Assessment Models (DREAM). Part 1: Model development and validation. Journal of Hydraulic Research, 44: 291–307.
13. Cui YT, Braudrick C, Dietrich WE, Cluer B, Parker G (2006b). Dam Removal Express Assessment Models (DREAM). Part 2: Sample runs/sensitivity tests. Journal of Hydraulic Research, 44: 308–323.

14. Wall DH (Ed.) (2004). Sustaining Biodiversity and Ecosystem Services in Soils and Sediments. Island Press, Washington, DC.
15. Gob F, Houbrects G, Hiver JM, Petit F (2005). River dredging, channel dynamics and bedload transport in an incised meandering river (the River Semois, Belgium). River Research and Applications, 21: 791–804.
16. Pedersen ML, Friberg N, Larsen SE (2004). Physical habitat structure in Danish lowland streams. River Research and Applications, 20: 653–669.
17. www.environment-agency.gov.uk
18. Förstner U (1987). Sediment-associated contaminants – an overview of scientific bases for developing remedial options. Hydrobiologia, 149: 221–246.
19. MAFF and Welsh Office (1995). Shoreline Management Plans: a Guide for Coastal Defence Authorities. MAFF Publications, London.
20. Defra (2006). Shoreline Management Plan Guidance. Vol. 1 Aims and Requirements. Defra Publications, London.
21. Halcrow (2002). Futurecoast [CD-ROM], produced for Department of Environment, Food and Rural Affairs, UK. Further information can be found from http://www.defra.gov.uk/environ/fcd/futurecoast.htm.
22. Apitz S, White S (2003). A conceptual framework for river-basin-scale sediment management. Journal of Soils and Sediments, 3: 132–138.
23. Hohensinner S, Habersack H, Jungwirth M, Zauner G (2004). Reconstruction of the characteristics of a natural alluvial river-floodplain system and hydromorphological changes following human modifications: the Danube River (1812–1991). River Research and Applications, 20: 25–41.
24. Becvar M (2005). Estimating typical sediment concentration probability density functions for European rivers. Unpublished MSc thesis, Cranfield University, UK.
25. White S, Fredenham E, Worrall F (2005). Sediments in GREAT-ER. Estimating sediment concentration ranges for European Rivers. Final report. Cranfield University, UK.
26. Apitz SE, Power B (2002). From risk assessment to sediment management: an international perspective. Journal of Soils and Sediments, 2: 61–66.
27. Babut M, Oen A, Hollert H, Apitz SE, Heise S, White S (2007). Prioritisation at river basin scale, risk assessment at site-specific scale: suggested approaches. In: Heise S (Ed.), Sustainable Management of Sediment Resources: Sediment Risk Management and Communication. Elsevier, Amsterdam, pp. 107–149.
28. Macklin MG, Brewer PA, Hudson-Edwards KA, Bird G, Coulthard TJ, Dennis IA, Lechler PJ, Miller JR, Turner JN (2006). A geomorphological approach to the management of rivers contaminated by metal mining. Geomorphology, 79: 423–447.
29. Apitz SE, Carlon C, Oen A, White S (2007). Strategic framework for managing sediment risk at the basin and site-specific scale. In: Heise S (Ed.), Sustainable Management of Sediment Resources: Sediment Risk Management and Communication. Elsevier, Amsterdam, pp. 77–106.
30. Owens PN (2005). Conceptual models and budgets for sediment management at the river basin scale. Journal of Soils and Sediments, 5: 201–212.
31. Joziasse J, Heise S, Oen A, Ellen GJ, Gerrits L (2007). Sediment management objectives and risk indicators. In: Heise S (Ed.), Sustainable Management of Sediment Resources: Sediment Risk Management and Communication. Elsevier, Amsterdam, pp. 9–75.
32. MAFF (2001). Flood and Coastal Defence Appraisal Guidance. Overview (Including General Guidance). Ministry of Agriculture, Fisheries and Food, London.
33. Heise S (Ed.) (2007). Sustainable Management of Sediment Resources: Sediment Risk Management and Communication. Elsevier, Amsterdam.
34. European Environment Agency (2005). Conceptual Framework: How We Reason. European Environment Agency, Copenhagen.

35. Smeets E, Weterings R (1999). Environmental Indicators: Typology and Overview. European Environment Agency, Report no. 25, Copenhagen.
36. White S, Deeks L, Apitz SE, Holden A, Freeman M (2006). Desk based study of the sediment inputs to the Broads catchment, with the identification of key inputs and recommendations for further targeted research and management to minimise inputs. Final report: Phases I and II. Cranfield University, UK.
37. Gray A-M (2006). Correlation between chlorophyll a data and phytoplankton contribution to sediment. Unpublished MSc thesis, Cranfield University, UK.
38. Wakelin T, Kelly A (2007a). Sediment Management Strategy. Broads Authority, Norwich.
39. Wakelin T, Kelly A (2007b). Sediment Management Strategy: Action Plan 2007/08. Broads Authority, Norwich.

Sustainable Management of Sediment Resources: Sediment Management at the River Basin Scale
Edited by Philip N. Owens

Regulatory Frameworks for Sediment Management

Susan T. Casper

Environment Agency, 10 Warwick Road, Solihull, West Midlands B92 7HX, UK

[*Note: The opinions expressed are those of the author and do not necessarily represent those of the Environment Agency*]

1. Introduction

The aquatic environment has long been recognized as a highly valuable resource, necessary for a range of activities including transport, recreation, agriculture and other industry, flood defence, and for the preservation of important habitats and ecosystems. Protection of the aquatic environment has now made it essential for stakeholders to develop and apply principles of sustainable water management that take into account the full range of demands on this resource, including societal and economic pressures, as well as environmental ones. In this context, much of the recent legislation is aimed at providing a more integrated approach to balancing the needs of the ecosystem against the various anthropogenic pressures in order to develop sustainable management practices.

The SedNet Strategy Paper [1] describes sediments as having important ecological, social and economic value. It recognizes sediments are an integral part of river systems, and play a large part in determining the health of an aquatic ecosystem since both sediment quality and quantity issues may result in adverse impacts on the ecological status of rivers, lakes, wetlands, floodplains, estuaries and the marine environment. Sediment management therefore has to consider the economic removal and use or disposal of sediments, the *in-situ* influences of sediments on the health of the ecosystem, and appropriate assessment and management of contaminated sediment.

One issue that remains particularly problematic for integrated sediment management is the differing conceptualization of sediment as an environmental component amongst different stakeholders, from potential contaminant, to fundamental natural substrate, to siltation nuisance. This variation in viewpoint may, in part, result in the current regulatory situation wherein there is no specific sediment legislative framework, but rather sediment management is explicitly or implicitly included in other regulatory contexts. Unfortunately, this means that across Europe, with the exception of management of dredged material, there are few specific legislative requirements for sediment, and the level of consideration afforded to sediment management is primarily left to the discretion of individual countries. This obviously has an impact in hampering efforts to harmonize sediment management, both in EU Member States, and more widely across the countries of Europe.

As noted in Chapter 2 (this book), drivers for sediment management tend to operate at a local rather than basin scale, often as a result of its impact on some aspect of the waterbody, such as the need to maintain navigation, rather than for its own sake in terms of supply and transfer. 'Sediment' as a solid phase component of the lithosphere may be alternately described as earth, soil, sediment, silt, sand, clay, dredged material or waste, depending on the scientific context, regulatory framework or management objective ([2], see also Chapter 1, this book).

Table 1 gives some internationally accepted definitions of soil, sediment and dredged material, though the ways in which these are interconnected in a dynamic system must be recognized. For example, Köthe [3] notes that soils on river floodplains are composed primarily of deposited river sediments, and can therefore exhibit the same characteristics of contamination as channel sediments, but are under the scope of soil legislation. The definition and value of sediment may, therefore, vary according to its perceived function, and even its particular placement in the aquatic system, in both fresh and marine waters. In order to address most effectively the interconnecting issues of sediment management, it is necessary to operate at the waterbody scale (e.g. river basins, marine waterbodies).

This chapter will give an overview of existing and developing sediment-relevant legislation, and consider the influence of non-regulatory drivers. For information on monitoring, see Bergmann and Maass [2].

2. The regulatory perspective

Historically much legislation has been developed in response to specific issues, from the early nationally specific legislation such as the Alkali Acts in the UK to the EU Dangerous Substances Directive, Drinking Water Directive, Groundwater Directive, and Integrated Pollution Prevention and Control Directive, amongst many others. The majority of the European Community's legislation, developed in the mid 1970s and early 1980s, followed by a second wave in the early 1990s, was developed to address specific substances, sources, uses or processes, and tends to concern protection of the environment for

Table 1. International definitions of materials relevant to sediment management (source: adapted from [2])

Soil	ISO [4]	Upper layer of earth's crust composed of: mineral parts, organic substance, water, air, living matter.
	Dictionary of Physical Geography [5]	The material composed of mineral particles and organic remains that overlies the bedrock and supports the growth of rooted plants.
Sediment (fluvial)	Dictionary of Physical Geography [5]	Particles derived from rock or biological material that are, or have been, transported by water.
	SedNet [6]	Suspended or deposited solids, of mineral as well as organic nature, acting as a main component of a matrix, which has been or is susceptible to being transported by water.
	PIANC [7]	Material, such as sand, silt or clay, suspended in or settled on the bottom of a water body.
	WFD AMPS [8]	Particulate matter such as sand, silt, clay or organic matter that has been deposited on the bottom of a water body and is susceptible to being transported by water.
Contaminated sediment	SedNet [9]	Sediments that have accumulated hazardous (intrinsic physical/chemical activity) substances (as a result of anthropogenic activities).
Dredged material	OSPAR [10]	Sediments or rocks with associated water, organic matter etc., removed from areas that are normally or regularly covered by water, using dredging or other excavation equipment.
	ISO [11]	Materials excavated during, for example, maintenance, construction, reconstruction and extension measures from waters.
	London Convention [12]	Material dredged that is by nature similar to undisturbed sediments in inland and coastal waters.

anthropocentric use. In some cases, such as the water environment, this has led to a considerable number of individual pieces of legislation, some examples of which are given in Table 2. This gives an indication of the scope of the issues

Table 2. Selected examples of EU Directives in water legislation (source: adapted from [13])

Legislation	Main purpose
Dangerous Substances Directive 76/464/EEC	To control the release of dangerous substances to water.
Freshwater Fish Directive 78/659/EEC	To ensure water quality meets the requirements for a healthy freshwater fish population.
Shellfish Directive 79/923/EEC	Aims to ensure a suitable environment for the growth of shell fisheries, and water of good quality to reduce the risk of food poisoning.
Bathing Water Directive 76/160/EEC and Revised Bathing Water Directive 2006/7/EC	To protect the environment and public health, and maintain amenity use of designated bathing waters (fresh and saline) by reducing the risk of pollution. The revised legislation aims to protect bathers from microbiological health risks, and promotes water quality and management actions.
Groundwater Directive 80/86/EEC	To prohibit the direct and indirect discharge of List 1 substances to groundwater, and limit the discharges of List 2 substances.
Surface Water Abstraction Directive 75/440/EEC	To protect the quality of water intended for use as drinking water.
Urban Wastewater Treatment Directive 91/271/EEC	To prevent the environment from being adversely affected by the disposal of insufficiently treated urban waste water, and ensure all significant discharges are treated before discharge to inland surface waters, groundwaters, estuaries or coastal waters.
Nitrates Directive 91/676/EEC	Aims to reduce pollution of surface and groundwaters by nitrates from agricultural sources.

covered in this context, and has been recognized as a piecemeal and inconsistent approach, the effect of which has also been compounded by inconsistent implementation across the EU.

More recent legislative frameworks, such as the Water Framework Directive (2000/60/EC), provide a greater focus on protection of the environment for ecocentric reasons, and also aim to develop more integrated approaches to the variety of issues we face, both across policy areas and between Member States. In addition to this, there is a greater emphasis on aspects of scale, with local environmental management issues also needing to be integrated at the river basin scale. This presents a significant current challenge for sediment management, and highlights the need for a corresponding integration of the different aspects of legislation pertaining to sediment, as well as integration of legislation with management frameworks (see Chapters 2 and 8, this book).

2.1. Water

Sediment is largely considered a component of the aquatic environment, so much of the legislation relating most clearly to sediment comes from this field.

2.1.1. The Water Framework Directive

The Water Framework Directive (WFD) [14] is intended to provide an overall framework for the protection and regulation of water management in Europe, encompassing a holistic approach to water management through the concept of river basin planning, and an emphasis on qualitative (chemical and ecological) and quantitative status. The key objectives outlined in the WFD (Article 1) are to:

- prevent further deterioration, and protect and enhance the status of aquatic ecosystems and associated wetlands;
- promote sustainable water consumption; and
- contribute to mitigating the effects of floods and droughts.

Since the WFD aims to protect the physical and biological integrity of all aquatic ecosystems, and thereby provide a mechanism for sustainable human water use, integrated ecologically orientated assessment of surface water status, with corresponding objectives, is a central theme [15]. Overall, the Directive requires the achievement of 'good status', in all natural waters (for surface waters it is good chemical and ecological status), except where derogations are identified, and an integrated approach to the regulation of all surface waters, estuaries, coastal waters and groundwaters by eco-region. Good status is described as a 'slight deviation' from conditions existing or expected to exist in a situation of minimal human impact, and for groundwater requires that quantity and quality does not adversely impact surface water status or the ecology of groundwater-dependent terrestrial ecosystems [15].

When fully implemented, the WFD will supersede several of the older Directives (e.g. Surface Water Abstraction, Freshwater Fish, Shellfish, Groundwater and Dangerous Substances), and aid the implementation of others (e.g. IPPC, Urban Wastewater Treatment, Habitats and Nitrates).

Although the WFD does not specifically direct Member States to monitor sediment quality, there are a number of ways in which sediment quality could be included as a tool for the assessment of water quality and ecological status [16]. For example:

- as an integral component affecting the health of monitored benthic species;
- as an important pathway in the exposure of biota to chemical contaminants;
- as an environmental matrix to which good status can be applied as well as to water and biota; and
- as a matrix for which standards can be applied for chemical quality.

Thus the WFD provides a mechanism for sediment issues to be addressed within the regulatory and management framework in situations where sediment is identified as impacting ecological quality. However, it remains to be seen to what extent these provisions are employed by Member States in dealing with sediment issues at both the catchment and basin scale.

Although there is generally a lack of standards for assessing the chemical quality of sediments, and the development and implementation of such standards for sediment remains problematic, it is worth noting that a 'standstill' condition applies to sediments under the provisions of the Dangerous Substances Directive. This provides a limitation control on emissions of priority substances to aquatic systems, and a requirement for monitoring of sediment at relevant locations under the Directive. This will remain a requirement of the WFD when it fully supersedes the Dangerous Substances Directive, although there is currently no anticipation of the development of further sediment standards for at least the first round of river basin planning (2009–2015).

In the UK, sediment delivery has been recognized as a diffuse source pressure for surface waters by the WFD River Basin Characterisation process [17]. The initial characterization process has identified approximately 21% of waterbodies in England and Wales as 'at risk' or 'probably at risk' of failing ecological objectives due to sediment delivery pressures. This assessment is based on a risk characterization, which combines the risk of sediment delivery with an assessment of the sensitivity of receptors and high-risk land-use factors, to generate an overall risk scale [18]. However, current confidence in the outputs is fairly low due to uncontrolled errors in some of the data used, low spatial resolution and a lack of validation, and it is anticipated that the current sediment delivery pressures assessment will be revised in the second round of river basin characterization. In Scotland and Ireland, the characterization process also recognizes sediment delivery as a diffuse pressure, but has not assessed its impact separately to overall diffuse source pressures, so it is difficult to make any comparison on that basis. The characterization process also identifies morphological pressures on waterbodies (rivers, lakes, transitional and coastal waters), which will tend to have a relationship with sediment pressures and effects, and which generally demonstrate a higher level of adverse impacts among the waterbodies assessed [17]. Unfortunately, one significant issue that remains is that of the impact of contaminated sediments, which is currently not adequately represented in the identification of impacts and pressures process. Although this may be partly taken into account in the assessment of individual substances, and is context specific, this is likely to impede any development of appropriate sediment *quality* objectives.

The river basin characterization process also clearly demonstrates the link between land-based activities and potential impacts on the achievement of good status in waterbodies, through the identification of the various pressures and industry sector contributions. Catchment-sensitive farming initiatives in the UK (see Section 4.1), for example, provide a land management framework that recognizes the importance of land connectivity to waterbodies, and so enables land and water management to be integrated at the catchment scale. It can be used to address issues of soil erosion and sediment delivery, as well as runoff of other contaminants to watercourses, and so provides an effective mechanism for achieving good status in water bodies.

The Water Framework Directive also establishes a system of 'protected areas', which will be given particular protection because of their use (e.g. drinking water or fisheries), or because of the need to conserve habitats and species of importance that depend on water. It is expected that these specially identified areas will be given a high priority in terms of achieving WFD objectives for good status (as well as the objectives of the existing legislation). Thus the WFD will aid the implementation of directives such as the Birds Directive (79/409/EEC) and the Habitats Directive (92/43/EEC).

2.1.2. Nature protection relevant to sediment management

The aim of the Habitats and Birds Directives is the conservation of natural habitats of wild flora and fauna, and the preservation of sufficient diversity and areas of habitats for birds in order to maintain healthy populations of all species. This requires a series of measures to be implemented at designated sites in order to restore natural habitat types and wild species of Community interest to a 'favourable conservation status'. This includes assessment of factors that may influence the long-term distribution and abundance of species populations, and the distribution, structure and function of protected habitats. Special areas of conservation (SACs) and special protected areas (SPAs) are also classified under the Habitats and Birds Directives, respectively, as sites of European importance, and collectively make up the Natura 2000 EU network of protected areas [19].

The Directives place a duty on the competent authority to avoid any significant adverse impact that could lead to the deterioration of habitat or disturbance of species of European importance. The Habitats Directive Review of Consents process in the UK [20] requires assessment of 'no adverse effect on site integrity' and 'appropriate assessment' in relation to discharges to designated sites. This intrinsically rather than explicitly includes sediments. For example, contaminated sediments could reasonably be included as a factor that may influence the conservation status of an aquatic species or habitat of importance. Equally, the maintenance of the sedimentary environment has a significant role to play in conserving important habitat types such as salt marshes and tidal mud flats.

However, while the Habitats Directive takes a more modern approach to nature conservation than the earlier Birds Directive, they remain some of the weaker pieces of EU legislation, with a general lack of effective implementation in Member States, and are some of the most litigated Directives in the EU [19]. The Habitats Directive is also described as being weak in its lack of provisions for the marine environment, and by not obliging the designation of sites for the protection of migratory species [19]. In addition to this, many of the provisions of the Directive, such as the concept of 'favourable conservation status', are set out in guidelines that are not legally binding, and therefore do not assist in underpinning the aims of the Directive [19].

2.1.3. The marine environment

The marine environment can be complex in terms of environmental protection, because of the transboundary nature of pollution and control across national, territorial and international waters. Thus, there tends to be wider participation of the EU in various international conventions in this particular environmental arena, consisting primarily of protection of the marine environment, prevention of oil spills and harmonization of national programmes for pollution control [13].

The marine environment is also, of course, partly regulated and protected under such water legislation as the WFD (for transitional and coastal waters to one nautical mile from the baseline) and the Habitats Directive (for example, where SACs or SPAs are located in estuarine or coastal areas). In these cases, the implications for sediment regulation and management are as discussed above (see Sections 2.1.1 and 2.1.2). In some cases, there are specific objectives for sediment, for example standards for suspended sediment for shellfish and EC bathing waters under the WFD [21], but in most cases the sediment perspective mainly comes in the form of support for achievement of ecological objectives.

In addition to the general water legislation that applies to estuarine (transitional) and coastal waters, the more recent EU Marine Strategy [22], adopted as a first step towards a Marine Framework Directive [23], also specifically applies to the marine environment. In the UK, the Government Department for the Environment, Food and Rural Affairs (Defra) has published a consultation document [24] for a UK marine bill, which will set out the government's vision for the marine environment as clean, healthy, safe, productive and biologically diverse oceans and seas.

While sediments are not perhaps a clear target of such legislation, as in other cases, there is recognition of the implicit need to protect and conserve sediments within the broader ecological objectives that underpin the proposed approach to conservation, protection and management of the environment. For example, a technical paper on the development of marine ecosystem objectives [21] describes how sediments needed to support herring spawning grounds are protected from adverse impact due to drilling or dredging activities by virtue of restriction of these activities where site-specific surveys identify such grounds.

2.2. Soil

While a European Soil Framework Directive does not yet exist, the European Commission's EU Thematic Strategy for Soil Protection has just recently been adopted [25]. Five working groups were set up to help develop the soil thematic strategy, and examined three of the eight identified threats to soils: erosion, decline in soil organic matter and soil contamination. In addition, they examined two cross-cutting themes, namely monitoring, and research and development. The remaining threats to soil identified by the Commission include: soil sealing, soil compaction, decline in biodiversity, salinization, and floods and landslides [26]. It is expected that a forthcoming Communication from the European Commission will include an assessment of the current situation as well as an outline of the strategy's objectives, and details for an Extended Impact Assessment and a Soil Framework Directive addressing the key threats to soil outlined in the original proposal [26].

Sediments have been integrated into the soil strategy, although the way in which they will be managed has yet to become fully clear. In many ways this may parallel the WFD, with little specific guidance on the assessment and management of sediments, but with instruments in place to allow for the management of sediments where there may be issues of concern. Soil and sediment are connected through suspension in water, connectivity and transport pathways through the catchment, and deposition from the river channel onto the floodplain (see Chapter 4, this book). Sediment will, therefore, have some bearing on soil issues such as erosion, flooding and contamination, where contaminated sediments may be transported from the active channel onto river corridor soils, and soil management, particularly with respect to erosion, will therefore also support key aspects of sediment management [27].

Some European countries already have soil protection legislation in force, which may or may not specifically include sediments. For example, in the Netherlands, sediments (as subhydric soils) are included in the Dutch Soil Protection Act, whilst in Germany they are excluded [3]. This is relevant for sediment management in terms of the deposition of fine sediments on floodplains, the potential for these deposited sediments to carry associated contaminants and the requirements within soil legislation for intervention or remediation when certain levels of soil contamination are reached [3]. In this context, it is not easy to divorce sediment management from soil protection.

In the UK, Defra has published a First Soil Action Plan for England 2004–2006 [28]. The Soil Action Plan builds on an earlier Draft Soil Strategy for England (2001), and has been informed by European initiatives such as the WFD and the work on the European Thematic Strategy for Soil Protection [25]. It is anticipated that this Soil Action Plan will help to highlight UK priorities, and provide a framework for action on soil protection in conjunction with existing legislation, schemes and instruments, such as implementation of cross compliance within the Common Agricultural Policy reform, and the Environmental Stewardship scheme [26, 28]. The Soil Action Plan recognizes many of the linked soil–sediment issues, such as erosion, infilling of lakes, siltation effects in rivers, flood risk and potential contamination issues. This thereby provides a mechanism to link soil and sediment management with other legislative drivers such as the WFD, Nitrates Directive and Habitats Directive, which also have aspects pertinent to land management, soil protection, prevention of contamination and protection of biodiversity.

2.3. Waste

In effect, much of the legislation applying most explicitly to sediments is waste legislation as it relates to dredged material, partly because this has long been recognized as a specific sediment issue needing appropriate management. Sediment dredging also crosses the boundaries of human health in terms of disposal of contaminated dredged material, and socio-economics in terms of maintenance dredging programmes. These factors have undoubtedly raised the profile of sediment management in this particular arena, in comparison with other sediment issues perhaps currently less stringently regulated by EU legislation.

Dredged material can be described as sediment that has been targetedly removed from its original position in a water system, for three principle reasons:

1. Remediation dredging – for environmental reasons (e.g. water quality).
2. Maintenance dredging – for sediment profile reasons (navigation, water stage).
3. Capital dredging – for port development, pipeline/outfall placement and other construction reasons.

The issue of sediment dredging is split by the questions of 'when and where' to dredge for the different drivers (maintenance, remediation and capital), and 'what to do' with the resultant material. The question regarding when and where dredging should take place, and if appropriate how it should be compensated for in environmental terms, is usually addressed by legislation more generally pertaining to the aquatic environment or environmental assessment, such as the WFD, Habitats Directive and Environmental Impact Assessment Directive (97/11/EEC). Present EU waste legislation does not address intervention points for sediment dredging, rather it addresses the issue of appropriate disposal of the material, once dredged, in a manner ensuring it does not pose any further or new threat to human health or the environment. Thus 'cradle to grave' management of sediment dredging is also split by the legislative context.

2.3.1. Waste regulations for dredged sediments

The European Framework Directive on Waste [29], implemented by the national waste regulations of the Member States, is applicable to sediments if they comply with the definition of waste as 'any substance or object which the holder disposes of, or is required to dispose of, pursuant to the provisions of national law in force', and is independent of the quality of the sediment [3]. If the dredged sediments are defined as waste, they may be categorized by two sub-codes contained in the European Waste Catalogue [30], namely 17 05 05 (dredging spoil containing dangerous substances) and 17 05 06 (dredging spoil other than those mentioned in 17 05 05). Thus the waste classification of dredged material depends primarily on assessment of the hazardous nature of the waste, which can have significant financial implications. In practical terms, except cases where sediment has been dredged for remediation purposes as a result of significant contamination, waste dredged material would usually be classified as non-hazardous according to the chemical criteria of the European Waste Catalogue.

The European Landfill Directive [31] regulates waste landfilling, including dredged material if it is to be disposed of with no further use in view. This only applies, therefore, if sediments cannot be beneficially re-used, and contains specific exceptions in this context including:

- the spreading of sludges (including those from dredging operations) on agricultural land for improvement purposes;
- the use of inert waste suitable for redevelopment/restoration and filling-in work, or for construction work in landfills; and
- the deposit of non-hazardous dredged sludges alongside small waterways from where they have been dredged, or in surface waters provided that certain conditions are met in each case.

These exemptions can provide a mechanism to encourage beneficial re-use of sediment rather than disposal to landfill (for further information see Bortone and Palumbo [32]), but it is important to note that application of the exemption still requires that the activity will meet the relevant objectives of the Member State's national waste legislation framework. In the UK, for example, this means dredged sediment re-use should meet the requirements of the Waste Management Licensing Regulations (1994) [33] outlined in Schedule 4 of the regulations [23]. This includes 'ensuring that waste is recovered or disposed of without endangering human health, without using processes and methods which could harm the environment, and in particular without risk to water, air, soil, plants and animals; or causing nuisance through noise or odours; or adversely affecting the countryside or special places of interest'. However, in cases where a Food and Environment Protection Act (1985) licence is issued for an operation between the high and low water marks, the Waste Management Licensing Regulations (1994) exempts the licensed activity from the controls of the Environmental Protection Act (1990) [34].

2.3.2. Disposal of dredged sediment

Waste legislation follows the basic principles of (1) avoidance of waste, (2) beneficial use (including treatment) and (3) landfilling [3]. All three options are relevant in an integrated sediment management framework, although the application of guidelines in the decision-making process will tend to differ at the national level. It does mean, though, that if sediments are non-hazardous, and can remain environmentally sound in the aquatic environment, then the definition of waste does not apply, and the relevant water or waterway legislation should apply instead. For disposal into the marine environment, this would normally require licensing under the Food and Environment Protection Act (1985) [13], though there is an increasing drive for beneficial re-use rather than disposal (also see [32]).

The dredged material guidelines of several of the international conventions relating principally to the marine environment also demonstrate that the preferred option for sediment management, in practice, is to leave sediments in the water in an environmentally sound and sustainable way whenever possible [3].

Part of the key decision-making process in the management of dredged sediment, therefore, is the determination of whether the sediment is sufficiently contaminated to be unacceptable for disposal in the aquatic environment, or is considered relatively uncontaminated and can be disposed of in an environmentally acceptable way by relocation. It should be noted though that dredged sediment considered too contaminated for disposal in the aquatic environment is not necessarily or automatically classed as hazardous waste, and it is this middle ground that can prove difficult to assess. By virtue of the interested parties, national guidelines tend to provide assessment criteria for aquatic disposal (relocation or confinement) in inland waters, although there may be some issues of harmonization in relation to dredged material in some of the larger inland European basins. In contrast, those for coastal waters are more in line with the guidelines set out by the international conventions, such as the OSPAR countries action levels. In both cases there is a need for mechanisms to define the 'contamination status' of the sediments under consideration.

This can be dealt with under the European Hazardous Waste Directive (91/689/EEC), implemented through national legislation, which sets out a framework by which contaminated soils, sediments or dredged spoils can be designated as hazardous or non-hazardous, in order to determine appropriate disposal routes, e.g. to landfill. However, due to the characteristics and ecological roles of sediments in natural waters, there is a view that the application of waste regulations derived principally for dealing with 'man-made' wastes is not necessarily entirely appropriate for dredged sediment [2]. The development of sediment quality criteria (SQC), applicable in fresh and marine waters, supports decision-making in sediment management in providing quality indicator values to guide thinking on how sediment should or should not be dealt with. The problem is that these values have not been developed in a standardized way, but on the basis of national scientific considerations and objectives, so tend to vary between countries [2]. For this reason, and because of scientific uncertainties in the relationship between the presence of a substance and the causation of actual harm, the current perspective generally is that the use of these guideline values as pass/fail criteria is not necessarily an appropriate method of sediment quality assessment [35]. The alternative approach currently advocated is for the development of 'weight of evidence' frameworks, taking into account ecotoxicological, ecological as well as chemical factors, in determining level of contamination and risk to the environment (also see Barceló and Petrovic [36]). This marks the transition from hazard-based management, characterized by the evaluation of the presence of a contaminant, to risk-based management, characterized instead by the effect the contaminant is likely to have on a target receptor (also see Heise [37]).

Dredged sediment that is classified as sufficiently contaminated to be deemed hazardous waste may require some pre-treatment before it can be disposed of to landfill. In particular, the Landfill Directive requires no liquid waste is landfilled, meaning dredged sediment would, at the very least, probably need to be de-watered, with associated costs. This inevitably leads to increasing costs of disposal overall, and may prove a barrier to the remediation of some aquatic environments. A common alternative for dealing with contaminated dredged material considered to be of unacceptable quality for sea disposal is confined disposal [34], e.g. the Slufter depot near Rotterdam [38]. This is a dedicated landfill, often in the intertidal zone, in which the dredged sediment is kept wet and anoxic to prevent leaching of associated contaminants. However, these facilities require careful management as drying out of materials, or rainfall infiltration could result in contaminant leaching.

Clearly, dealing with the dredging and disposal of both contaminated and non-contaminated sediments remains a complex and difficult issue due to the variety of drivers, classifications and regulations pertaining to the final solution. For further information on the treatment of sediment and dredged material see Bortone and Palumbo [32].

3. International marine and freshwater management

As noted in Section 2.1.3, the majority of international environmental conventions relate to the marine rather than the freshwater environment. This is primarily a result of common ownership issues and the transboundary nature of pollution in this sector of the environment, although there is clearly a similar issue for large European river basins.

3.1. International maritime conventions

Table 3 gives an overview of some of the existing conventions and their purposes. It is worth noting those that apply globally to marine areas, for example the London Convention, and those that have a more restricted area of application, such as the Helsinki Convention. While the globally applied conventions have a considerable scope for influencing appropriate environmental management, the more areally restricted conventions have a greater opportunity to reflect the interests of the bordering nations, and may also therefore be more persuasive in attracting signatories.

Sediment management in coastal areas is often largely concerned with dredged material management, since large volumes of sediment in estuaries are

Table 3. Some global and regional international maritime conventions (source: based on [2])

Convention	Area of application	Purpose
London [12]	Global marine areas	Prevention of marine pollution by dumping of wastes and other matters (1972)
		1996 protocol to the London Convention (came into force March 2006)
		Dredged material assessment framework (DMAF) adopted in 1995
MARPOL [39]	Global marine areas	Prevention of pollution from ships (1973)
OPRC [40]	Global marine areas	Prevention of oil pollution (1990)
		A protocol to this convention (HNS Protocol) deals with marine pollution by hazardous and noxious substances
OSPAR [10]	NE Atlantic, including the North Sea	Guidelines for the management of dredged material (1998, and revised 2004)
Helsinki [41]	Baltic Sea	Dredged material guidelines (1994)
Barcelona [42]	Mediterranean Sea	Protection of the Sea against pollution

commonly dredged for navigation purposes, and the material disposed of either to land or the aquatic environment. A significant element of these international maritime conventions deals with dredged material assessment and management, although there are also conventions dealing specifically with the disposal of wastes and potentially toxic materials from ships. The International Convention for the Prevention of Pollution from Ships (MARPOL), for example, covers pollution of marine waters by oil, chemicals, packaging, sewage and other refuse from ships, while the International Convention on Oil Pollution Preparedness, Response and Co-operation (OPRC) provides a framework for international cooperation in respect of major marine incidents or threats.

Of critical importance in the management of dredged material is the sediment quality: whether it can be regarded as relatively 'clean' sediment, or contains concentrations of potentially damaging contaminants. Several European countries use a relative scale of sediment quality criteria with upper and lower 'action' values for contaminant concentrations to aid management decision-making [2]. For the reasons outlined in Section 2.3.2 on dredged sediment disposal, the use of these values as pass/fail criteria is often not recommended for sediment assessment, and most of the conventions recommend methods based on a 'weight of evidence' approach [35].

In the UK, the OSPAR convention is the primary driver for the National Marine Monitoring Programme (NMMP), comprising a network of sites in UK coastal waters monitored for sediment as well as water quality criteria. This is an important network, as it is one of the few programmes that supports comprehensive sediment sampling and monitoring in the UK. If we also consider estuarine and coastal areas as the ultimate sink for catchment sediments (see Chapter 1, this book), this may also provide a potential indicator of sediment issues higher up the transport network.

3.2. International maritime organizations

There are a number of international organizations working towards a common understanding of the issues affecting marine waters, and appropriate management and protection, particularly of sensitive marine areas. It is not the intention to provide a comprehensive listing here, but to highlight some of the principal associations relevant to sediment and dredged material management as follows:

- *International Maritime Organization (IMO) [43]*
  Deals with environmental protection from marine pollution, and acts as the Secretariat for the London Convention/Protocol.
- *Permanent International Association of Navigation Congresses (PIANC) [44]*
  Recommendations and frameworks for the management of dredged material.
- *Central Dredging Association (CEDA) [45] and International Association of Dredging Companies (IADC)*
  General guidance on dredged material management.
- *Dutch–German Exchange on dredged material (DGE) [46]*
  A platform for exchange of information on dredged material management and related issues.
  Aims to elaborate the state of the art situation in relevant disciplines (e.g. legislation, ecotoxicology, chemistry and treatment) in sediment and dredged material management.
  The DGE initiative was extended in 2005 to DGE+ (Multilateral Exchange on dredged material) with the introduction of partners from the UK, France and Belgium.
- *European Dredging Association (EUDA) [47]*
  Regulatory guidance for the disposal of dredged material in Europe.

Further information on these organizations and their recommendations can be found in material by Bergmann and Maass [2] and Köthe [3] and websites referenced above.

3.3. River basin conventions

Just as the transboundary nature of pollution in the marine environment requires international cooperation and principles for effective management, the same is true of the larger European river basins, which may cross several European states, each with their own administrative structures and interpretations of the legislation. The Convention on the Protection and Use of Transboundary Watercourses and International Lakes [48] provides a framework for promoting cooperation between riparian states in a catchment, and developing guidance for transboundary water management, although it does not provide recommendations for specific river basins.

Subsequently, many European countries have successfully cooperated in the environmental management of large rivers (see Bergmann and Maass [2] for major European International Commissions for the protection of transboundary rivers and lakes), but the strategies and action plans developed have tended to focus on pollution abatement, and improvement of the ecological status of waters, rather than on sediment management [2]. This is probably not surprising, as it essentially reflects water legislation such as the WFD in placing little specific focus on sediment management, though it may eventually raise questions on approaches to large-scale sediment management. The main exception to this is the International Commission for the Protection of the River Rhine (ICPR), which has established agreement on the disposal and relocation of dredged material – though this still does not account for the full role of sediment in environmental and ecological improvement through erosion, deposition, contamination (sink and source) and habitat [2]. Work of the International Commission for the Protection of the Danube River (ICPDR) is described in Chapter 8 (this book).

4. Initiatives and non-legislative drivers: a UK perspective

The general lack of specific legislative drivers for sediment does not necessarily prevent adequate management of sediment issues at a local level, since all Member States will interpret the existing legislation in ways that will enable achievement of appropriate environmental objectives. Some of the environmental initiatives that are responses to the legislation may, therefore, encompass management of sediment issues, even if the original legislation was not explicit in this context. Additionally, local or regional non-regulatory drivers may also address sediment in the context of issue-based responses. These can be particularly useful in providing frameworks to deal with local, practical, 'ground level' problems, and form an issue-based perspective to management that can complement the focus of legislation and regulatory drivers which tend more towards management of the environmental compartment.

The following text aims to provide examples of some of the current initiatives and drivers that exist in the UK to address practical environmental management at river and catchment scale. As with the legislation, for most of these sediment management is not the primary reason for their initiation, but sediment issues are nevertheless encompassed within them.

4.1. Catchment Sensitive Farming Programme

The Catchment Sensitive Farming (CSF) Programme takes forward the UK Government's strategic review of diffuse water pollution from agriculture, and includes action under the Nitrates Directive, as well as developing measures to meet WFD requirements [49]. It aims to provide a pro-active approach to agricultural diffuse pollution, by reducing sources of this pollution in river catchments through land management practices, in order to ensure emissions to water are consistent with WFD ecological requirements [50].

For example, the England Catchment Sensitive Farming Initiative (which started in April 2006) aims to promote voluntary action by farmers in 40 priority catchments across England to tackle the problems of diffuse water pollution from agriculture. A joint initiative between Defra, the Environment Agency and the Natural England Partnership will provide a network of 40 catchment sensitive farming officers, coordinated at river basin district scale, to advise farmers on ways to tackle water pollution. It will focus at the local level and bring farmers together with advisors, conservation bodies, water companies and other stakeholders to effect environmental improvement [49, 50].

The Environment Agency and the former English Nature (now Natural England) together identified the priority catchments based on data gathered for the WFD risk assessment purposes on nitrates, phosphorus and sediment pollution, together with data on sensitive freshwater fisheries, chalk streams, failing bathing waters, groundwaters and SAC designated lakes. The English Nature prioritization of designated sites at risk of diffuse water pollution from agriculture was also taken into account in this process [50]. Clearly, while sediment is not the only 'culprit' in agriculturally based diffuse water pollution issues, nor necessarily the largest, it may well act as a vector in transporting other, possibly better recognized, pollutants such as phosphorus, as well as acting on its own to contribute to turbidity and smothering effects (see Chapter 1, this book). It is, therefore, not only an appropriate target for management, but is one that potentially can be significantly improved by changes in land management practices that reduce soil erosion within the catchment. In this way, the CSF initiative can be considered to contribute to sediment management by reducing soil erosion and the consequent smothering effects on riverbed habitats, to reducing water pollution by associated pollutants for WFD objectives, and also to soil protection objectives under the Thematic Strategy for Soil Protection.

4.2. Salmon Action Plans

Fish are an essential component of healthy aquatic ecosystems and an indicator of environmental quality in aquatic environments. Maintenance of fish stocks and the diversity of wild fish are therefore an important part of the UK's commitment to international conventions such as those on biodiversity, sustainable development and the conservation of salmon, as well as the EU Freshwater Fish Directive. The Environment Agency (in England and Wales) supports the development of both Fisheries Action Plans for all catchments, and Salmon Action Plans for the 62 principal salmon rivers in England and Wales [49].

Salmon Action Plans are the mechanism by which the Environment Agency meets the objectives of its National Salmon Management Strategy at a local level. It provides, among other things, specific spawning targets for the principal salmon rivers in England and Wales, against which stock and fishery performance are assessed [49]. The Defra website on fisheries management [51] describes this as providing 'a more objective approach than has been previously applied to salmon management in England and Wales', and which has been advocated by the North Atlantic Salmon Conservation Organization (NASCO) to facilitate salmon management in the international context.

Each plan contains a variety of actions and objectives, which are designed to increase spawning and nursery areas, improve water quality and protect habitats, reduce the impact of bird predation on young salmon, improve stock assessment, reduce exploitation and minimize obstructions to migration [50].

Since sediment intrusion has a significant impact on the development of salmon spawning redds through siltation and smothering effects ([52], also see Chapter 1, this book), sediment management in terms of erosion and delivery is a key factor in the environmental improvement of salmon rivers, which has long been recognized.

4.3. Sustainable Urban Drainage Systems

The adoption and development of Sustainable Urban Drainage Systems (SUDS) for dealing with the drainage of water from land have several well-recognized benefits [50] including:

- reducing flood risk;
- minimizing diffuse pollution;
- maintaining or restoring natural flow regimes; and
- improving water resources and enhancing amenity.

While the principle objective of SUDS is local management of water resources, particularly in the context of increasing urbanization and climate change effects on the hydrological regime, there is also a large associated element concerned with water quality, amenity (biodiversity) and diffuse pollution. Although the agricultural element of diffuse pollution often warrants much attention, this should not obscure the role that urban diffuse pollution can also play in environmental degradation. Urban and road runoff of contaminants, along with preferential transport of sediment through drains and ditches and along the highways network, can make a substantial contribution to the load received by the receptor waterway (also see Chapter 4, this book).

SUDS, therefore, in developing systems to intercept and attenuate runoff of water and sediment, can make an important contribution to the management and control of urban diffuse pollution. The Scottish Environment Protection Agency (SEPA) website [53] currently provides additional information on SUDS research and monitoring, and the activities of the SUDS Monitoring Project Steering Group, which has, among others, members from SEPA, the Environment Agency, and the Northern Ireland Environment and Heritage Service (EHS).

4.4. The UK Biodiversity Action Plan

The UK Biodiversity Action Plan (UKBAP) was published by the Government in 1994 in response to Article 6 of the Convention on Biological Diversity held in Rio de Janerio in 1992 [49]. The aim of the UKBAP is to develop national strategies for the conservation of biological diversity and the sustainable use of biological resources, and includes contributions from the Government, statutory conservation agencies, academia and the voluntary sector [49]. The UKBAP lists species and habitats identified for conservation, and assigns various actions to different organizations, some of which have consequently developed their own more specific action plans, usually at a local or regional level [50].

The national strategy for the implementation of the UKBAP is translated into action at the local level through Local Biodiversity Action Plans, with coordination across the UK undertaken by the UK Biodiversity Partnership (recognizing that following devolution, biodiversity conservation policies primarily reside with each of the four countries) [49]. Again, while sediment may not be explicitly mentioned here at the higher strategic level, it is inextricably bound up with the maintenance of good habitats, prevention of diffuse pollution impacts and land management strategies for the conservation of key species at the local level, and must therefore be given due consideration.

4.5. Beneficial use of dredged sediment

Section 2.3.2 noted that the disposal of dredged sediment into the marine environment requires licensing under the Food and Environment Protection Act (1985) [13], and that under this Act there is also a requirement for the licensing authority to consider alternative beneficial options for dealing with this material. Within the concept of beneficial re-use of dredged sediment, there is therefore a driver not only to deal with a potential waste disposal issue, but also to develop schemes for environmental (or other) improvement. In this case, the dredged material is regarded as a resource rather than a waste material, which through appropriate placement can be used to derive environmental benefits such as the re-creation of lost habitat. This is particularly important in marine and coastal areas, which have some of the most ecologically sensitive and important habitats, such as special areas of conservation and special protected areas.

The Determination of the Ecological Consequences of Dredged Material Emplacement (DECODE) group represents organizations with interests in the utilization of fine-grained dredged material for environmental enhancement, habitat sustainability and flood defence [54]. The group currently comprises members representing many fields of interest, from marine ecological research to environmental regulation, and aims to assist in the harmonization and knowledge transfer of research on beneficial use of dredged sediment.

Some of the advantages of these beneficial use schemes highlighted by the DECODE group include [54]:

- Flood and coastal defence – studies demonstrating the role of saltmarshes in protecting vulnerable coastlines and sea defences, and the options for using dredged material to protect or create saltmarsh where these are currently being eroded.

- Sediment cell maintenance – where in-estuary placement of dredged material can be used to reduce erosion of inter-tidal banks and saltmarshes, and minimize the perturbations to the estuary's natural sediment cell maintenance that might otherwise occur through essential dredging (i.e. a re-balance of the net sediment erosion and deposition).
- Habitat conservation/enhancement – protection of eroding and/or creation of new saltmarshes and mudflats as habitat for important invertebrate and bird species. These habitats are eventually capable of functioning like natural systems, and help to mitigate the loss of these valuable environments in the UK due to the impacts of erosion and reclamation.

5. Sediment management

Environmental management within the regulatory framework has historically depended largely on the assessment of environmental quality by measurement against standards, and the application of instruments such as consents for discharges to prevent pollution, and tools for remediation to deal with unacceptable levels of contamination. Thus appropriate monitoring and assessment regimes are necessary to underpin effective management, such as those currently used to manage water quality.

Unfortunately, in the case of sediment management, the general lack of comprehensive monitoring schemes tends to result in an ad-hoc localized approach to decision-making, and the lack of widely accepted sediment standards can complicate assessment. Both the OSPAR and London conventions called for the development of sediment quality criteria to assess dredged material prior to disposal, but these have mainly been developed on a national basis, and consequently can vary considerably from country to country, although there is some cooperation amongst countries involved in River Basin Commission programmes [2]. Other standards, such as the Canadian sediment quality guidelines for fresh and marine waters [55], have been similarly developed, and although they may be accepted as appropriate guidelines across several countries, there is currently no common sediment standard within the international regulatory framework. The WFD seeks to develop some standards for sediment, for example for the priority substances. However, this is complicated by the increasingly held view [36, 37] that a weight of evidence approach is more appropriate than simple pass/fail criteria, and that any decision-making should be placed in the context of a risk assessment framework, taking into account principles of sustainability. Detailed information on risk assessment approaches across Europe is given in Den Besten [56].

Whilst there are current sediment *quality* standards available, even if not internationally harmonized, it is worth noting that no standards currently exist for sediment *quantity* management. This in particular makes it very difficult to deal with well-recognized issues arising from sediment delivery (or erosion) effects, such as the silting of spawning gravels. Initiatives such as Sustainable Urban Drainage Systems can include provisions to manage sediment delivery, for example through building regulations, but it is difficult to make this effective when there is a lack of consensus on what constitutes acceptable and unacceptable delivery. Thus, for sediment management to achieve the legislative aims of, for example, good ecological status under the WFD, there is a need for increased monitoring, improved assessment tools and an ability to link these to effective measures for improvement taking socio-economic drivers into account.

6. Summary and conclusions

This chapter has highlighted a number of legislative and policy provisions that exist for appropriate management of sediment quality and quantity issues, but also suggests the disparate way in which they tend to be applied could be a significant factor in failing to achieve the fullest possible environmental benefit.

For example, the Dangerous Substances Directive 'standstill' provision for specific substance concentration in sediment applies to monitoring at Dangerous Substance Directive regulated sites, which may be relatively few and far between, and have often arisen as a response to point source inputs from industry. This may not, therefore, capture the issue of diffuse sediment contamination, which whilst possibly of lower specific concentrations in any particular given location, could be of considerable geographic extent (see Chapter 4, this book).

The Habitats Directive also has some requirement for consideration of sediment, but only where it may be considered to have the potential to affect 'site integrity', and only in relation to designated sites (e.g. SACs, SPAs). If sediment contamination is found to be a relevant issue, the Habitats Directive does contain powers for regulators to review and potentially amend the discharge consents of industries discharging to the designated area in order to reduce the inputs of damaging contaminants. However, in order for this to be effective, it is necessary to have a full understanding of the transport routes of the contaminants to the sites, and an ability to identify the specific sources of the contaminants, which may prove difficult. In this respect, the methods described in Chapters 4 and 5 (this book) offer considerable potential.

Waste regulation with respect to dredging issues also applies at very specific sites, and principally applies in respect of disposal options rather than the need to dredge. Dredging of contaminated sediments only occurs where other legislation highlights this as a problem needing to be addressed, or when sediments dredged for maintenance purposes are found to contain sufficiently high levels of specific substances that they have to be designated as contaminated. Maintenance dredging of 'clean' sediments often requires less stringent management, with the main issue being a disposal option that will not cause the receiving habitat to be adversely impacted.

Thus, much of the legislation relating to management of sediment applies at restricted locations, whereas the reality is that the major sediment issues of both quality and quantity in the most part stem from diffuse sources and impacts. The WFD is a clear step forward in its recognition of the diffuse impact issue of sediment delivery, and its aim to integrate the different drivers in management of the aquatic environment, but there is still some distance to go in achieving effective integrated sediment management under its auspices.

Sediment has an intrinsic role in the provision of good ecologically sound habitat to support a wide variety of species, and therefore also in maintaining good ecological quality. However, while much of the most recently developed legislation is driven fundamentally by an ecologically orientated focus, the role and influence of sediment in supporting, for example, the achievement of good hydromorphological, ecological and chemical status in waterbodies for the WFD is not always fully recognized. Herein also lies a problem in the relation of sediment-based contributing factors to legislative drivers where sediment is not a specific consideration. For example, the extent to which pathogens in sediments may contribute to impacts on bathing waters may not be fully recognized through water-column monitoring for the Bathing Water Directive.

It must also be remembered though that there is a wider context to be considered within the regulatory perspective, in that for effective *regulation* there must be: an identifiable problem; a cost-effective solution; and preferably a responsible 'owner' to effect the necessary changes. So, there is a scientific issue that has to be addressed in furthering our current understanding of the role and contribution of sediment and sediment-associated contaminants to perceived ecological or environmental issues; that is, the link between sediment delivery, and any associated contamination, and actual environmental impacts (also see Chapter 1, this book). There is a difficulty in that sediment issues tend to be diffuse issues, and the scope of such issues can mean their full resolution is prohibitively expensive to the stakeholder community. Finally, it is often very difficult to identify parties responsible for its damage, particularly in cases where the historical record of contamination is a confounding factor, which can make it difficult to recover costs for restoration work. The Environmental Liability Directive [57] could play a part in this, in that it aims to establish a framework to prevent significant environmental damage, or to rectify damage after it has occurred, by requiring operators to pay prevention and remediation/restoration costs for such damage. However, it is not yet entirely clear how this may work in relation to sediment, or if it can be practically applied in this respect, given the difficulties outlined above.

In conclusion, there is a clear need for greater legislative recognition of the role and influence of sediment quality and quantity issues in supporting habitats and the achievement of ecological objectives, and fully integrated sediment basin management to tie together the diverse interests of the basin stakeholders. A sediment management framework (or series of frameworks) needs to be developed to integrate, clarify, support and further the legislation and regulations that currently exist for sediment management.

Acknowledgements

I would like to thank Helge Bergmann, Vera Maass, Harald Köthe and Sjoerd Hoornstra for their contributions to this chapter, and Peter Howsam and Chris Vivian for reviewing an earlier version and providing helpful comments and suggestions.

References

1. SedNet (2006). Strategy Paper. http://www.SedNet.org
2. Bergmann H, Maass V (2007). Sediment regulations and monitoring programmes in Europe. In: Heise S (Ed.), Sustainable Management of Sediment Resources: Sediment Risk Management and Communication. Elsevier, Amsterdam, pp. 207–231.
3. Köthe H (2003). Existing sediment management guidelines: an overview. Journal of Soils and Sediments, 3: 144–162.
4. ISO (1996). ISO 11074-1, Soil quality – Vocabulary terms and definitions relating to the protection and pollution of the soil.
5. Thomas DSG, Goudie A (2000). The Dictionary of Physical Geography. Blackwell, Oxford.
6. SedNet (2007). Glossary. In: Heise S (Ed.), Sustainable Management of Sediment Resources: Sediment Risk Management and Communication. Elsevier, Amsterdam, pp. 269-274.
7. PIANC (2002). Environmental guidelines for aquatic, nearshore and upland confined disposal facilities for contaminated dredged material. PIANC International Navigation Association, Working Group EnviCom 5 report.
8. WFD AMPS (2004). Discussion document on sediment monitoring guidance for the EU Water Framework Directive. Working Group on Analysis and Monitoring of Priority Substances, Subgroup on Sediment.
9. Salomons W, Brils J (Eds) (2004). Contaminated Sediments in European River Basins. European Sediment Research Network, SedNet. TNO, Den Helder, The Netherlands.
10. OSPAR (2004). Revised OSPAR guidelines for the management of dredged material (http://www.ospar.org).

11. ISO (1999). ISO 11074-3, Soil quality – Vocabulary terms and definitions relating to soil and site assessment.
12. London Convention (2000). Specific guidelines for the assessment of dredged material (http://www.londonconvention.org).
13. NSCA (2000). Pollution Handbook. National Society for Clean Air and Environmental Protection, Brighton.
14. EC (2000). Directive 2000/60/EC of the European Parliament, and of the Council establishing a framework for the Community action in the field of water policy.
15. Scheuer S (2005). Water. In: EU Environmental Policy Handbook. A Critical Analysis of EU Environmental Legislation. European Environmental Bureau, Brussels.
16. Powers E (2001). Sediments in England and Wales: Nature and Extent of the Issues. Environment Agency, Bristol.
17. Environment Agency Water Framework Directive, and River Basin Characterisation pages (http://www.environment-agency.gov.uk).
18. Environment Agency WFD River Basin Characterisation: Technical assessment method for sediment delivery to rivers. River Basin Characterisation pages (http://www.environment-agency.gov.uk and http://www.environment-agency.gov.uk/commondata/acrobat/r_sediment_t_v1_1008046.pdf).
19. Falter C, Scheuer S (2005). Nature. In: EU Environmental Policy Handbook. A Critical Analysis of EU Environmental Legislation. European Environmental Bureau, Brussels.
20. Environment Agency (2001). EU Habitats Directive and Regulations Process Handbook for Agency Permissions and Activities. Environment Agency, Bristol (http://www.environment-agency.gov.uk).
21. Rogers SI, Tasker M, Whitmee D (2005). A technical paper to support the development of marine ecosystem objectives for the UK (http://www.defra.gov.uk/corporate/consult/marinebill/techpaper.pdf).
22. EC (2005a). Thematic strategy on the protection and conservation of the marine environment. Communication from the Commission to the Council and European Parliament. COM(2005)504.
23. EC (2005b). Framework for Community action in the field of marine environmental policy (Marine Strategy Directive). Proposal for a Directive from the European Commission COM(2005)505.
24. http://www.defra.gov.uk
25. http://ec.europa.eu/environment/soil/index.htm
26. http://www.defra.gov.uk/environment/land/soil/europe/index.htm
27. Quevauviller Q, Olazabal C (2003). Links between the Water Framework Directive, the Thematic Strategy on Soil Protection, and research trends with focus on pollution issues. Journal of Soils and Sediments, 3: 243–244.
28. Defra (2004). The First Soil Action Plan for England: 2004–2006. Department for Environment, Food and Rural Affairs (Defra), London.
29. EC (1975). European Framework Directive on Waste, Council Directive of 15 July 1975 on waste (75/442/EEC); modified by Decisions 91/156/EEC and 96/350/EC.
30. EC (2000). European Waste Catalogue 2000/532/EC.
31. EC (1999). European Landfill Directive 1999/31/EC.
32. Bortone G, Palumbo L (Eds) (2007). Sustainable Management of Sediment Resources: Sediment and Dredged Material Treatment. Elsevier, Amsterdam.
33. Waste Management Licensing Regulations (1994). And amendments thereof.
34. Vivian C (2006). Personal communication.

35. Delvalls TA, Andres A, Belzunce MJ, Buceta JL, Casado-Martinez MC, Castro R, Riba I, Viguri JR, Blasco J (2004). Chemical and ecotoxicological guidelines for managing disposal of dredged material. Trends in Analytical Chemistry, 23: 819–828.
36. Barceló D, Petrovic M (Eds) (2007). Sustainable Management of Sediment Resources: Sediment Quality and Impact Assessment of Pollutants. Elsevier, Amsterdam.
37. Heise S (Ed.) (2007). Sustainable Management of Sediment Resources: Sediment Risk Management and Communication. Elsevier, Amsterdam.
38. http://www.slufter.com/nl/homepage/index.jsp
39. International Convention for the Prevention of Pollution from Ships (1973).
40. International Convention on Oil Pollution Preparedness, Response and Co-operation (1990). (Entered into force 13 May 1995).
41. Convention on the Protection of the Marine Environment of the Baltic Sea Area (1992). (Entered into force 17 Jan 2000).
42. Convention for the Protection of the Mediterranean Sea Against Pollution (1976). (Entered into force 12 Feb 1978).
43. International Maritime Organization (http://www.imo.org).
44. Permanent International Association of Navigation Congresses (http://www.pianc-aipcn.org).
45. Central Dredging Association (http://www.dredging.org).
46. Dutch–German Exchange on dredged material (http://www.htg-baggergut.de).
47. European Dredging Association (http://www.european-dredging.info).
48. United Nations Economic Commission for Europe (UNECE) (1992). Convention on the Protection and Use of Transboundary Watercourses and International Lakes (Water Convention).
49. Department of Environment, Food and Rural Affairs (http://www.defra.gov.uk).
50. Environment Agency (http://www.environment-agency.gov.uk).
51. http://www.defra.gov.uk/fish/freshwater/fishman
52. Environment Agency (2006). Water Resources Habitats Directive Technical Advisory Group: Appropriate Assessments for Riverine SACs. Environment Agency, Bristol (http://www.environment-agency.gov.uk).
53. Scottish Environment Protection Agency (SEPA) (http://www.sepa.uk/dpi/suds).
54. http://www.cefas.co.uk/decode
55. Canadian Council of Ministers of the Environment (CCME) (1999). Canadian sediment quality guidelines for the protection of aquatic life: summary tables. In: Canadian Environmental Quality Guidelines. CCME, Winnipeg.
56. Den Besten PJ (2007). Risk assessment approaches in European Countries. In: Heise S (Ed.), Sustainable Management of Sediment Resources: Sediment Risk Management and Communication. Elsevier, Amsterdam, pp. 153–205.
57. Environmental Liability Directive (2004). 2004/35/CE – 21 April 2004.

Edited by Philip N. Owens

Sediment and Contaminant Sources and Transfers in River Basins

Kevin G. Taylor[a], Philip N. Owens[b], Ramon J. Batalla[c] and Celso Garcia[d]

[a]*Department of Environmental and Geographical Sciences, Manchester Metropolitan University, Manchester M1 5GD, UK*
[b]*National Soil Resources Institute, Cranfield University, North Wyke Research Station, Okehampton, Devon EX20 2SB, UK. Now at: Environmental Science Program, University of Northern British Columbia, 3333 University Way, Prince George, British Columbia, V2N 4Z9, Canada*
[c]*Department of Soil and Environmental Sciences, University of Lleida/Forestry and Technology Center of Catalonia, Av. Alcalde Rovira Roure, 191, E-25198 Lleida, Catalonia, Spain*
[d]*Department of Earth Sciences, University of the Balearic Islands, Edifici Beatriu de Pinós, Ctra. de Valldemossa, E-07122 Palma de Mallorca, Spain*

1. Introduction

In recognizing the need to assemble information on the sediment–contaminant system so as to inform sediment management frameworks and plans (Chapter 2, this book), this chapter describes the main sources and pathways of sediment (Section 2) and contaminants (Section 3) in river systems. Information on sediment sources and pathways is needed because source control and/or controlling the pathways by which sediments are delivered to watercourses is often the optimal management option. The sources of sediment and contaminants initially are discussed separately because in many situations their sources are different in terms of type (soil, sewage treatment works), form (particulate, in solution) and spatial location (forests, urban). Section 4 then describes the transfer pathways and processes that control sediment–contaminant fluxes through river systems from source to sink. Finally, Section 5 describes some of the main perturbations to sediment and contaminant sources and transfers. Further information on sediment–contaminant conceptual models and budgets which link sources, pathways and fluxes is contained in Chapter 1 (this book), while Chapter 5 (this book) provides information on the methods and tools used to assemble information on sources and transfers.

2. Sources and pathways of sediment to river channels

This section considers the sources and pathways of sediment found in river basins. Sources of contaminants and nutrients, both in particulate form (i.e. bound to sediment) and in solution, are discussed in Section 3. Here, source is defined as the physical origin, location or type of the sediment (e.g. topsoil, bank, road dust, etc.), whereas pathway is used to describe the delivery process or mechanism (e.g. overland flow, road network, subsurface tile drain). Mass movements events are defined here as a source, although they are also a delivery process.

It seems appropriate at the outset to recognize that sediment creation, mobilization and transfer are a natural part of river system behaviour. Weathering of rock by physical, chemical and biological processes, and the actions of wind and rain as mobilizing and transporting agents, mean that particles of mineral and organic material move under the influence of gravity from higher to lower elevations. In many respects, these natural processes control sediment sources and transfers in river basins, and have been occurring over millennia. Over time, what has changed mainly has been the precise location and magnitude of these sources, although in recent centuries society has also introduced some new sources of sediment (e.g. mining wastes and solids from sewage treatment works (STWs) and combined sewer overflows (CSOs)) and drastically altered the rates at which sources and pathways supply sediment to the river system (see Section 5, and Chapter 1, this book).

In most river basins, the main 'natural' sources of sediment *to rivers* are:

- atmospheric dust deposition and wind erosion;
- mass movement events (such as landslides, debris flows, etc.); and
- erosion of soils by water.

The relative importance of these vary greatly from country to country [1], from basin to basin and within basins. Thus mass movement events are rare for most lowland landscapes (e.g. the Netherlands and lowland reaches of a basin), but may be the dominant source of sediment in steep mountain environments (e.g. Nepal and upland reaches of a basin).

Within the *river corridor*, additional sources of sediment include:

- erosion of channel banks by lateral channel migration (here defined as part of the river channel as it marks the boundary of the river channel and lies within its floodplain);

- erosion of floodplain deposits (by overbank flows during flooding events, as distinct from channel bank erosion);
- resuspension of channel bed sediment (fine- and coarse-grained) due to changes in flow condition; and
- erosion from cliffs and coastal areas (especially relevant in the lower parts of the river basin below the tidal limit).

Further, relatively recent sediment sources due to anthropogenic activities include:

- geological mines;
- construction sites;
- urban road network;
- in-stream gravel mining; and
- mineral and organic material from point sources such as STWs and industrial point discharges.

The following sections consider each of the sources in turn and the mechanisms and processes that control how they are delivered to watercourses. In Section 3, these sources of sediment are also discussed as potential sources of sediment-associated contaminants and nutrients, although for the latter type it seems more appropriate to group them according to the main land use that causes the sediment to be contaminated (i.e. agricultural, urban, industrial, etc.).

2.1. Atmospheric dust deposition and wind erosion

The deposition of sediment from atmospheric sources, such as atmospheric dust, is generally of limited importance in terms of the overall sediment budget of river basins. However, for environments that have limited vegetation cover, such as arid, semiarid, alpine and subalpine areas, there can be a significant amount of erosion of soil and fine exposed sediment (moraines, talus, etc.) by wind processes, which is subsequently deposited either within or outside of the basin from where it was derived [2, 3]. Reported annual rates of continental dust deposition vary from <10 g m^{-2} to about 200 g m^{-2} [4].

In the agricultural areas of Europe, wind erosion and sediment redistribution is generally of limited extent and magnitude compared to water erosion in these areas, or compared to wind erosion in arid and alpine areas. Consequently, rates of atmospheric dust deposition are low. The most extensive and most severe areas of wind erosion are in southeastern Europe, such as Romania, the Ukraine and Russia, although moderate wind erosion has also been reported for the Czech Republic and parts of France, the Netherlands and the UK [1].

Sediment supplied to river channels derived from atmospheric sources is usually fine-grained (here defined as <63 μm), although some wind-transported sediment derived from local sources may be within the range >63 μm to <1 mm. For atmospheric sediment derived from pristine environments, such as arid and alpine areas, the sediment will be relatively 'clean', whereas that derived from wind erosion in agricultural or urban areas (e.g. industrial emissions, opencast mining spoils) may be heavily contaminated (see Section 3.2).

2.2. Mass movements

Mass movement events can deliver large inputs of both fine-grained and, particularly, coarse-grained (>63 μm) sediment (Figure 1). These events are caused by an instability in the soil, surficial and/or bedrock material, such that the force to move due to loading and gravity is greater than the inherent resistance or strength of the material. Mass movements are often triggered by extreme precipitation or tectonic (i.e. earthquakes) events. There are many different types of mass movements, including landslides, debris flows, debris slides and rock avalanches (for further details see [5, 6]). Most mass movement forms (i.e. landslides) are rapid and provide an instantaneous pulse of material either directly to river channels or to areas near to channels, which are subsequently mobilized by erosion processes. Other mass movement forms, such as those associated with solifluction and gelifluction, are much slower and provide a more continuous supply of sediment. Mass movements are generally restricted to mountainous and upland environments, and thus represent an important, if not the dominant, source of sediment in these areas. In virtually all cases, the sediment supplied to channels from mass movement events has little or no contamination associated with it unless the location is a contaminated one, such as a mine site.

2.3. Erosion of soils by water and tillage operations

There are a variety of mechanisms that cause soil particles to be detached and redistributed, which include those due to wind (see Section 2.1), water, tillage, animals and harvesting of crops. Of these, water erosion processes (e.g. rill, inter-rill and gully processes) are generally dominant [7], particularly in terms of supplying sediment to river channels. The depth to which erosion occurs is temporally and spatially variable. In most situations it is the top few centimetres that are eroded, although when overland flow becomes concentrated, the depth of erosion can be much greater, creating rills and gullies (Figure 1).

In most agricultural areas, soil erosion by water represents the main source of the sediment transported in rivers. Numerous studies from around the world have used sediment fingerprinting and tracing techniques to determine the sources of fluvial sediment [8–12] (see Chapter 5, this book, for further details on these techniques). In most situations, erosion of topsoil (top 30 cm) represents the dominant sediment source. Thus, for example, Collins *et al.* [13] estimated that erosion of topsoil from communal cultivation, bush grazing and commercial cultivation contributed 64%, 17% and 2% of the sediment load at the outlet of the Kaleya catchment (63 km^2) in Zambia, with the remainder derived from bank erosion and some gully erosion. Similarly, Walling *et al.* [14] determined the main sources of the sediment transported in the lower reaches of the River Ouse, England, and several of its main tributaries. They found that typically between 60% and 85% was derived from topsoil, mainly from uncultivated land such as pasture and moorland (Table 1).

Tillage operations can also represent an important mechanism for the downslope movement of topsoil towards rivers [15, 16]. Although tillage operations may not actually deliver sediment into river channels *per se*, they represent an important process for mobilizing sediment and often result in a net downslope movement of sediment, which may, in turn, be delivered to channels by overland flow and/or wind erosion.

Although erosion of topsoil usually represents the main source of sediment in rivers in agricultural basins, there are a large number of pathways by which the sediment is delivered to the river. These include both surface (e.g. overland flow, gullies, etc.) and subsurface (e.g. through macro-pores and tile drains) pathways. Of particular note is the road and track network [17]. Important here

Table 1. Contributions from topsoil and channel bank material to the suspended sediment load transported in the downstream reaches of the River Ouse, England, and its main tributaries based on sediment samples collected between November 1994 and February 1997 (source: modified from [14])

River	Area	Source type contributions (%)			
	(km^2)	Woodland topsoil	Uncultivated topsoil	Cultivated topsoil	Channel bank material
Swale	1350	0	42	30	28
Ure	914	1	45	17	37
Nidd	484	7	75	3	15
Wharfe	818	4	70	4	22
Ouse	3315	0	25	38	37

is the fact that not only does the road and track network within rural and agricultural areas represent an important source of sediment due to erosion by water, machinery and grazing animals, but it also represents an efficient pathway for transfers of sediment to the river system. We are, however, lacking information on the modification of sediment properties and characteristics within the road/track network during transport and delivery processes.

2.4. Bank erosion

The erosion of channel bank material is a major source of sediment for most rivers. Material eroded from channel banks is important because it is delivered directly into the river channel and thus often provides an immediate impact (Figure 1), and because the sediment is often coarse-grained, thereby representing an important source of bedload material (see Section 4.1.2). Work in the UK has demonstrated that typically between 10% and 40% of the suspended sediment load of rivers in agricultural basins may be derived from channel bank sources [8, 9, 12, 14]. The relative contribution depends on a variety of factors including climate (rainfall and temperature), river flows (peak discharge and duration), properties of the bank material (e.g. particle size, organic matter content) and land use. For example, Carter *et al.* [18] determined the contribution of channel bank material along the length of the River Aire in England. In rural, headwater areas the contribution from channel banks was between 45% and 85%, whereas the contribution in the downstream urbanized reaches was <35%. This downstream decrease reflected the increased importance of other sources, such as road dust, and the fact that many of the channel banks in the urban areas were protected from erosion.

2.5. Erosion of floodplain deposits

Although floodplains represent a net sink of sediment over long time periods, it is important to recognize that at times they are also a source of sediment. During overbank flows sediment can be eroded from the surface of the floodplain (i.e. as different from lateral erosion of channel banks). There is very little information on the importance of floodplains as a source of sediment but several studies have suggested that it may be significant during periods of overbank flows. Information on floodplains as sources of contaminated sediment is contained in Section 3.7.

Figure 1. Sources of sediment to river channels (clockwise from top left): mass movement event, Canada; soil erosion on cultivated land, England; in-channel gravel and sand mining, Belize: and channel bank erosion, England (all photos by P.N. Owens)

2.6. Resuspension of in-channel sediment

Within the river channel itself, sediment can also be supplied to the overlying river water through the resuspension of in-channel sediment sources on the channel bed. In many respects such sediment represents temporary storage as part of the sediment transport continuum from initial source (i.e. soils) to sink (i.e. oceans), as opposed to a distinct primary sediment source. However, it is important to distinguish channel bed sediment from other sediment sources because it is activated only during certain hydrological conditions. In-channel sediment sources are usually activated when there is an increase in discharge, which in turn means that there is an increase in bed shear which resuspends the sediment (see Section 4.1 for further details). The amount and particle size composition of this sediment is dependent on discharge and other in-stream properties, but during extreme flow conditions it is possible to resuspend the fine sediment (i.e. silt and clay, <63 μm) stored on, and within, the surface layers of the channel bed and also erode coarser bed material (i.e. sand and gravel, >63 μm).

Studies have suggested that the amount of fine-grained sediment stored on and within the upper layers of the channel bed, and thus available for resuspension and incorporation within the water column as suspended sediment, varies from <1% to >16% of the annual suspended sediment flux [19, 20]. However, for individual storm events with high-flow conditions and large amounts of stored fine sediment, sediment contributions from channel bed sources may be considerably more significant and may represent an important source, particularly during the early stages of the hydrograph (i.e. until this source is exhausted). The sediment resuspended from the channel bed is also important to sediment (and contaminant) fluxes in rivers because this sediment is often different from that derived from other sources due to in-channel processes. Thus sediment from this source is often more flocculated (see Chapter 1, this book) and contains more organic material than sediment from topsoil and channel bank sources [21–23].

2.7. Coastal zone sources

As the EU Water Framework Directive considers estuarine and near-coastal zone (within 1 km of the coast) environments within the definition of a river basin, it is also important to consider sediment sources below the tidal limit of rivers. In many river systems, a significant proportion of the sediment load in tidal reaches and estuaries is derived from estuarine and coastal zone sources. Such sources include tidal mud flats and coastal cliffs, in addition to sediment derived from ocean sources. For the Humber estuary in England, for example, most of the sediment is derived from the marine environment and not from the contributing catchment ([24]; Chapter 1, this book) and this situation is likely to be true for many of the harbours and estuaries bordering the North Sea [25].

2.8. Anthropogenic sediment sources

Sources of sediment primarily associated with anthropogenic activities are mainly discussed in Section 3 in relation to sources of contaminants, because for many of these sources the sediment is contaminated due to the anthropogenic activities. Thus sediment derived from STWs, CSOs, industrial point sources, the urban road network and geological mining sites are usually enriched with metals, persistent organic pollutants, radionuclides, etc. Some anthropogenic activities, however, are important sources of essentially 'clean' sediment, and these include in-stream gravel mining (Figure 1). The latter is considered in more detail in Section 5 as a specific example of a disturbance to the 'natural' sediment system.

3. Sources and pathways of contaminants and nutrients to sediments

Here, for the sake of simplicity and in the context of this book, a contaminant is defined as a substance (natural or artificial) that, at a level above a threshold, increases the risk of a detrimental effect on the functioning of aquatic ecosystems and/or human health. Contaminants in river basins take a variety of forms, including metals, inorganic elements, organic compounds and radionuclides, and the major sources of these contaminants are highlighted in Table 2. It is important to be aware that many of these contaminants can be sourced from natural processes as well as anthropogenic activities, although in most cases anthropogenic inputs tend to dominate. A nutrient is an element that is essential for the growth of organisms. The most important nutrients in river basins are nitrogen (N) and phosphorus (P). Nitrogen is predominantly present as a dissolved phase, whilst over 90% of P is sediment-bound [26] and is, therefore, the nutrient of most concern from the viewpoint of sediment management. High levels of P have a detrimental impact upon river basins [26], and thus P is considered in this chapter.

Table 2. Examples of sources of sediment and associated contaminants to river basins

Material	Sources
Sediment (organic and inorganic)	Erosion from rural, agricultural and forested land, channel banks, urban road dust and construction, STW solids, atmospheric deposition, inputs from tidal areas and coastal zone (during flood and ebb tidal cycle)
Metals and metalloids (Ag, Cd, Cu, Co, Cr, Hg, Ni, Pb, Sb, Sn, Zn, As)	Geology, mining, industry, acid rock drainage, sewage treatment, urban runoff
Nutrients (P, N)	Agricultural and urban runoff, wastewater and sewage treatment
Organic compounds (pesticides, herbicides, hydrocarbons, PCBs, PAHs, dioxins)	Agriculture, industry, sewage, landfill, urban runoff, combustion
Xenobiotica and antibiotics	Sewage treatment works, industry, agriculture
Radionuclides (^{137}Cs, ^{129}I, ^{239}Pu, ^{230}Th, ^{99}Tc)	Nuclear power industry, military, geology, agriculture (secondary source)

In order to determine if sediment is contaminated, baseline and threshold-effect information are needed. In the case of artificial compounds (e.g. pesticides, PCBs, some radionuclides) the baseline value is zero, and contamination assessment is relatively straightforward. Pre-industrial historical values need to be established for elements with both natural and anthropogenic sources (e.g. heavy metals, P). In rare cases this can be accomplished through analysis of monitoring data or archived samples, but in most cases values are determined through the analysis of sediments accumulated through time [27]. Threshold-effect values are often determined through an assessment of the physical, chemical or biological nature of the sediment, or a combination of these. Increasingly, the end result is a series of threshold-effect values that collectively form Sediment Quality Guidelines, or other similar measures (e.g. critical level, critical load, weight-of-evidence assessments, etc.). For further information on the analysis and assessment of sediment see Barceló and Petrovic [28].

3.1. Types of contaminant sources

Contaminant sources may be of particulate, dissolved or gaseous form. The form they take will govern the pathways and transfers of the contaminants in the river basin. It has been clearly documented that for most metals the particulate form is dominant [29–31]. Up to 90% of metals are present in the particulate form, but this can vary from metal to metal, and between source types. Similar observations have been made for organic contaminants, P and radionuclides [32]. Whilst contaminant sources are predominantly particulate, there are important exceptions. Contaminants from sewage treatment works (e.g. P, Zn) can be predominantly in solute form, and metals from acid mine drainage are also in dissolved form due to the low pH of the waters. These dissolved contaminants, however, commonly become associated with the particulate phase, via mineral precipitation or surface adsorption, as solution concentration, pH and redox state change through mixing with dilute river water. Atmospheric forms can be important (e.g. Pb, Zn, dioxins) in combustion sources.

Contaminant sources may take one of two general forms – point sources and non-point sources – each of which requires specific approaches regarding identification and management. The quality of the sediment itself is determined by the input of substances from both point and diffuse sources. These sources contribute to the natural background level (e.g. erosion for nutrients and heavy metals) and elevated levels of organic chemicals, nutrients and heavy metals. Point sources are those sources originating from a single location, and as such are often readily identified. Thus, point sources are *identifiable points* and are (fairly) *steady in flow and quality* (within the temporal scale of years). The magnitude of pollution from point sources is not influenced significantly by the magnitude of meteorological factors, although there may be some exceptions such as certain operations of STWs and CSOs. Furthermore, such sources are generally easily controlled and monitored. Examples of point sources of contaminants relevant to sediment management include mines and mine waste, landfill sites, industries, STWs, CSOs and bedrock mineralization.

Non-point (or diffuse) sources of contaminants are those originating from a wide area. Diffuse pollution can be defined as: 'pollution arising from land-use activities (urban and rural) that are dispersed across a catchment or subcatchment, and do not arise as a process industrial effluent, municipal sewage effluent, deep mine or farm effluent discharge' [33]. Thus, diffuse sources are *highly dynamic spatial pollution sources* and their magnitude is *closely related to meteorological factors* such as precipitation. Such contamination is not manageable until the original sources are controlled [34].

As a result, the identification, monitoring and, in particular, the control of these diffuse sources present much more of a challenge to sediment management at the river basin scale. However, given the high level of success in controlling point sources of pollution (in part due to the success of the EU, River Commissions, national and regional directives and legislation – see Chapter 3, this book), these diffuse sources are now recognized as requiring the most effort for identification and control [35]. Examples of diffuse sources of pollution relevant to sediments in river basins include direct atmospheric deposition, urban runoff and sediments (i.e. from the road network), agricultural runoff and sediments (i.e. from soil erosion), the reworking of floodplain sediments (i.e. by bank erosion), historic contaminated sediments and background geology (also see Table 2).

Figure 2 presents the various diffuse and point sources and pathways contributing to the input of substances in a river system (also see Figure 6, Chapter 5, this book). Both point and diffuse sources contribute to the total contaminant load of rivers. A distinction between them is necessary for future restoration actions and determining the effect of past control measures at industrial sources.

3.2. Atmospheric deposition

Many contaminants are delivered to the Earth's surface by wet and dry atmospheric deposition. The original source of the contaminants can be natural (e.g. salts) or artificial (e.g. industrial discharges into the atmosphere). A good example of the latter is the release into the atmosphere of radionuclides such as Cs and Pu due to the Chernobyl incident in 1986 (a point source) and the

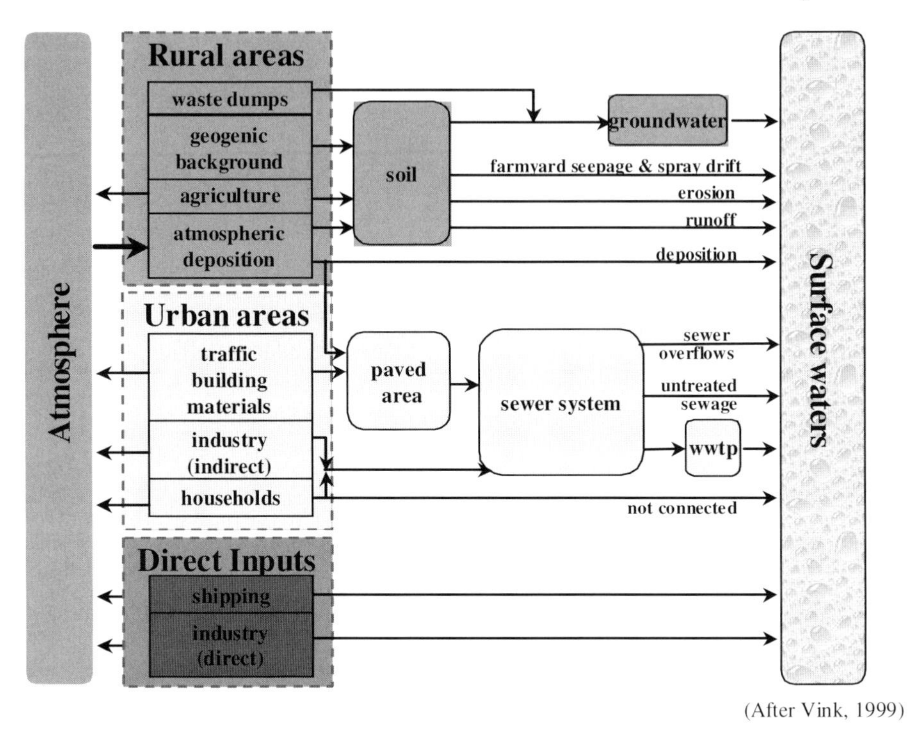

Figure 2. Flow of materials considered in a river basin (source: [36])

subsequent deposition and contamination of a large part of Europe, Turkey and the former Soviet Union [37]. Given the relative surface area of the land within river basins (typically >90%) compared to that of surface waters such as rivers, lakes and reservoirs, most atmospheric deposition is on to the land surface. The mechanisms by which the land-deposited contaminants are subsequently transferred and delivered to waters then depends on specific processes and pathways associated with each land use (agricultural, forested or urban), and these have been described above for sediment (Section 2) and are discussed in the sections below in terms of sediment–contaminant transfers. The direct deposition of contaminants onto the surface of water bodies is, however, important for several reasons. First, the contaminant is delivered direct to the water. Thus, there are virtually no, if any, opportunities for buffering between the point of deposition and waters, as might occur in soils, for example. Second, management of atmospheric deposition of contaminants is very difficult, as virtually the only option is the control of artificial discharges into the atmosphere from the original source: there are few viable intermediate options.

3.3. Bedrock mineralization and mining

Bedrock mineralization is a natural source of contamination to sediments and is almost always diffuse in nature. Individual areas of mineralization may act as point sources, and this has been utilized for mineral prospecting using stream sediment surveys (e.g. [38, 39]). However, more commonly bedrock mineralization is regional in nature and natural weathering processes provide metal-rich sediment to river basins. This can be quantified using base-line stream sediment surveys (e.g. the British Geological Survey G-BASE Mapping programme [40]). Such data are very important in river basins as they allow background data to be determined, which is an important aspect of river sediment management. High elemental levels in river basin sediment are not always a result of anthropogenic sources.

Mining has been documented to be a significant anthropogenic source of contamination. Sediment-bound contaminants enter river systems either from point (e.g. tailings effluent and other mine discharges) or diffuse sources (e.g. remobilization of contaminated alluvium) [41, 42]. Mining-sourced contaminants are mostly restricted to metals and inorganic contaminants such as As and SO_4^{2-}. Such contamination may be acute in nature, where contaminant releases are single events. The best documented recent example of such an event is that of the Aznalcollar Zn mine, southern Spain, in 1998, where a failure of a tailings impoundment led to the release of metal-rich sediment into the Guadalquivir River system with acute ecological effects [43]. It is more common, however, for contaminant releases to be longer-term, either through repeated discharges (e.g. regular tailings effluent discharge into the Río Pilcomayo, Bolivia [44]), or longer term dispersal of tailings or waste piles. In addition to particulate inputs, acid mine drainage (AMD) can be a significant source of dissolved metal contaminants. As these contaminants mix and dilute with river water, pH increases and precipitation of iron- and metal-oxides leads to sediment-bound contamination (e.g. [45]).

3.4. Urban sources

Urban environments are highly engineered environments, with major sources of anthropogenic contaminants. Urbanization gives rise to unique and specific sources of contaminants, many of which are not found in more natural river basins. Contaminants in urbanized environments are derived from a complex mixture of point sources (i.e. STWs and CSOs) and sources which are more diffuse (i.e. urban road network). The diffuse sources of contaminants in urbanized areas are numerous and include vehicle fuel combustion, emissions from catalytic convertors, tyre wear, vegetative plant fragments, garden soil, metallic fragments, concrete and cement, de-icing salt, and building material [46–48]. In addition, urban river basin sediments receive high organic matter and nutrient (N, P) point source inputs from STWs and CSOs [49–51], and also receive enhanced metal and organic contaminant species from urban industrial processes [52, 53].

Atmospheric contaminant sources are also important in urban environments. Lead, predominantly derived from leaded fuel (where tetraethyl-lead is used as an additive), has been of major concern. However, with the widespread reduction in use of leaded fuel, Pb levels in urban river systems are falling [54–56]. The platinum group elements (PGEs) platinum, palladium and rhodium are a relatively recent contaminant, having been emitted from urban environments since the early 1990s. PGEs act as catalysts in catalytic converters, and with the phasing-out of leaded fuel are currently the metals of most concern emitted from vehicle exhausts. There is evidence from many studies for the widespread dispersion and accumulation of PGEs in urban river basins and air-borne particulates [48, 57, 58].

There is a whole suite of organic contaminants (so-called persistent organic pollutants, POPs) sourced to river basins from urban environments that often become associated with sediment in the river system. These include PAHs (polyaromatic hydrocarbons), PCBs (polychlorinated biphenyls), hydrocarbons, dioxins, pesticides and herbicides. The sources of these are various and include both atmospheric and land-based sources. For example, PAHs in Vancouver, Canada, were observed to be sourced from biomass burning and vehicular emissions [59]. Probably, the largest source of organic pollutants derives from vehicular activity. Many of these are found in petrol or diesel (including benzene, toluene, naphthalene, PAHs), or associated with automobiles (including ethylene glycol, hydraulic fluids, styrene, oil lubricants). Pesticides and herbicides are applied to urban soils in residential areas and gardens, where they can be removed by runoff or erosion and sourced to river basin sediments.

3.4.1. STWs and CSOs: examples of urban point sources

Within the urbanized environment, STWs and CSOs are point sources of contamination that have had a major impact upon river basin quality in the past. However, due to water quality and discharge regulations, many of these sources are now controlled within acceptable levels, and evidence suggests that the contaminant levels of fluvial sediments from these sources have tended to decrease over the last few decades or longer [55]. However, it is worth noting that most of our understanding of the impact of these sources is based around water quality issues – very little data exists on the impact upon sediments of these sources. Recent research has shown that these point sources, especially in

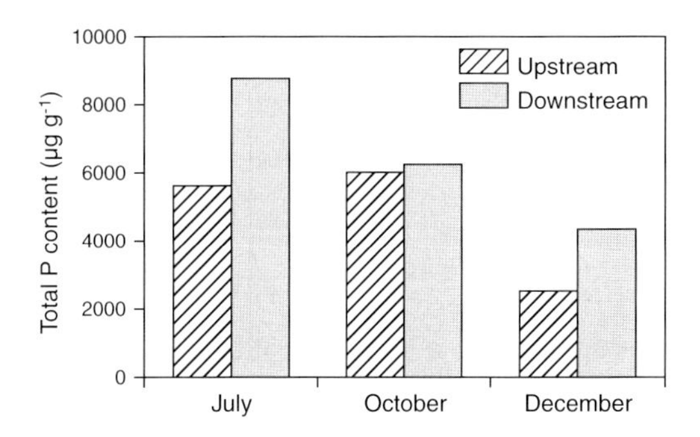

Figure 3. Phosphorus content of suspended sediment immediately upstream and downstream of a large STW, River Aire, England. There were no significant sources of sediment or phosphorus between the two sampling sites other than the STW (source: based on data in [51])

highly urbanized river basins, still have a major impact upon sediment-associated contaminant concentrations and fluxes [60]. For example, Owens and Walling [51] documented a clear increase in sediment-bound total-P (a major contributor to river eutrophication) as a result of urbanization in the River Aire, England. They documented changes in total-P from <2000 μg g^{-1} in upstream sections to >7000 μg g^{-1} (with values for individual suspended sediment samples >12,000 μg g^{-1}) in sections downstream of major urban areas. They concluded that this increase represented point inputs of P from STWs and CSOs, probably in dissolved form which subsequently sorbed onto the passing sediment (Figure 3). The input of these point sources was further supported by a change in the inorganic P to organic P ratio from <2 in upstream reaches to >4 in reaches downstream of urban areas, including the cities of Bradford, Halifax and Leeds. This is significant in that inorganic P is more bioavailable than organic P, further highlighting the impact of STWs and CSOs on sediments. Downstream increases in the contaminant content of fluvial sediment due to point source inputs was also documented in this basin for other contaminants, such as metals (Cr, Cu, Pb and Zn) and PCBs [60–63].

3.4.2. The urban road network: an example of an urban diffuse source

Urban sources of contamination are a major, and growing, diffuse source to river basins. One of the main diffuse sources of sediment and contaminants in urban areas is road dust (sediment source), which supplies material via road drains (sediment pathway), which, in turn, often discharge directly into rivers. Generally, there is limited quantitative information on the contribution from the road network to sediment and contaminant fluxes in urban river basins. A recent exception is that of Carter *et al.* [18] for the River Aire, England. Through the

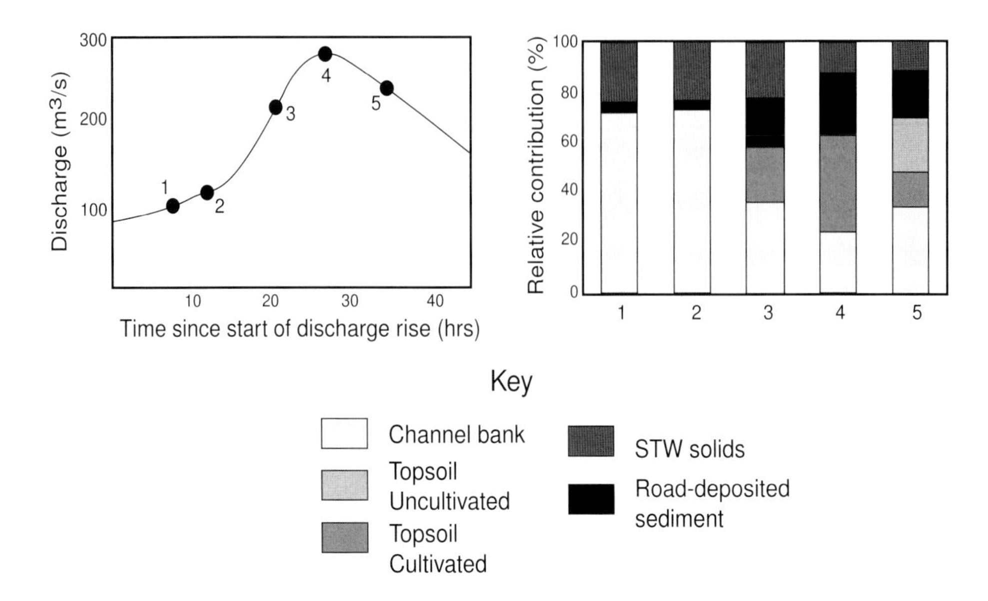

Figure 4. Variation in the relative contribution from surface material from uncultivated and cultivated areas, channel bank material, road dust and solids from STWs to five samples of suspended sediment collected from the downstream reaches of the River Aire, England, during a storm event during March 1998 (source: modified from [18])

application of statistically verified fingerprinting techniques (for further details of the technique see Chapter 5, this book), they showed that approximately 20% of the contaminated sediment flux in the urban river reaches was derived from the road network. Interestingly, this study also demonstrated that the contribution from this source increased during a storm event, reflecting the increase in the connectivity of the road network to the channel system as the storm progressed (i.e. more runoff), while the contribution from STWs and CSOs decreased during the hydrograph, reflecting dilution effects (Figure 4).

3.5. Industry

Industrial sources are those derived from manufacturing processes and power generation. Whilst a large range of contaminants can be derived from industrial processes, the most significant to river basins are metals, synthetic organic compounds (e.g. the organochlorine pesticides and PCBs) and radionuclides. Of metals, those with the most ecological impacts are Cd, Cr, Hg and Pb. Numerous studies have shown the impact that these sources can have on river basin sediments. For example, Walling *et al.* [62] showed that in the Aire–

Calder river basin in England, Cr in suspended sediment increased from around 100 μg g^{-1} in non-urban reaches to approximately 400 μg g^{-1} in urbanized reaches, while PCB congener concentrations showed a four-fold increase.

Radionuclides have been input into river basins via nuclear power generation and weapons testing. Such inputs may be widespread (e.g. ^{137}Cs from atmospheric weapons testing or Chernobyl fallout) or more localized (power generation). Nuclear power stations, due to their requirements, are commonly sited on coasts. Estuarine parts of river basins may, therefore, receive significant inputs of radionuclides (e.g. ^{137}Cs, ^{99}Tc, ^{129}I, ^{230}Th). These species are overwhelmingly associated with the particulate phase and can accumulate to significant levels in impacted river basins.

3.6. Agriculture

Agriculture probably represents the largest diffuse source of contaminants to rivers and water bodies in Europe. A variety of contaminants are delivered to waters from agricultural land and these include pathogens, metals, radionuclides, nutrients, pesticides (including herbicides, fungicides and insecticides) and other micro-organic contaminants [31, 32, 64–70] (Table 2). Most of these are either primarily sediment-associated or may become sorbed onto sediment within the aquatic environment. Thus, the application of fertilizers and pesticides to agricultural land results in both solute and particulate-bound inputs of N, P and pesticides to rivers. Of these, P and pesticides have the greatest affinity to the particulate phase [32, 64] and, therefore, are amongst the most important agricultural contaminant sources to sediments in river basins.

Many of the contaminants listed above do not occur naturally within the soil (such as ^{137}Cs – which is derived from the atom-bomb tests and the Chernobyl incident) or are present in soils at elevated concentrations due to atmospheric deposition, applications of wastes (e.g. sewage sludges or biowastes) or artificial inputs associated with farming practices (such as the application of P-based fertilizers). Most of the contaminants listed in Table 2 tend to be elevated in the surface layers of the soil profile: often in the upper 5 cm. This reflects either atmospheric deposition (as in the case of fallout radionuclides) or artificial inputs to the soil surface (as in the case of P), and the fact that many chemicals are sediment-associated and thus sorb tightly to soil particles (both mineral and organic) in the top layers [67, 71, 72]. Surface erosion processes (see Section 2.3) then export the contaminated sediment to the river system. Certain land management operations such as ploughing may, however, alter the depth distribution of contaminants and this will have an effect on the surface concentration, and thus the delivery of contaminants to waters due to soil erosion [67, 71, 72].

Although most contaminants associated with agricultural land are delivered to waters by surface pathways (i.e. overland flow, surface erosion and the road network; see Section 2.3), recent studies [17, 67, 68, 73, 74] have shown that there are also subsurface pathways (such as macropores and artificial drains) by which sediments and contaminants (both particulate and dissolved form) move through the soil and are delivered to surface waters (and groundwaters).

In agricultural areas, bank erosion also represents an important, although often neglected, source of contaminated sediment to rivers. Thus, Laubel *et al.* [75], for example, describe the importance of bank erosion as a source of P to streams in Denmark. As described in Section 2.4, one of the reasons for the importance of bank erosion as a source of contaminated sediment is that the delivery ratio is effectively 100% and material is delivered directly in to the channel system. Bank erosion and other channel sources of sediment are discussed further below.

3.7. Channel sources of contaminants

An issue of special importance is the 'historic' contamination of sediments as 'sleeping' sources of contamination in river basins. As new inputs of contaminants will continue to decrease due to the effects of existing and forthcoming environmental legislation (see Chapter 3, this book), the contribution of 'historically' contaminated sediments to contamination loads in river basins will gain in relative importance. This process is governed by re-erosion during high water discharges, by relocation of dredged material stemming from weirs and locks, and related retention and loss processes. For sediment management purposes, and for the evaluation of potential risks associated with accumulated contaminated sediments in river basins, these 'historic' contaminated sediments should be assessed.

Floodplains are sites of sediment accumulation within river basins and, therefore, are classically considered to be contaminant sinks [42, 76, 77] (Table 3), thereby preserving good temporal records of contaminant input [27, 55, 78]. These sinks of contaminants, however, can also become sources as a result of post-depositional processes, both chemical and physical. Contaminant elements in floodplain sediments can be affected by secondary chemical remobilization and the formation of new contaminant element-bearing minerals. Gäbler [79] showed that in overbank sediments contaminated by mining activities in the Harz Mountains, Germany, primary metal sulphide minerals within the overbank sediments were converted into other species in which the metal contaminants had higher mobility than in the original sulphides. Hudson-

Table 3. Estimates of the deposition and conveyances losses of sediment and associated contaminants on the floodplains bordering the main channels of the Rivers Swale (1346 km^2) and Aire (1002 km^2), England (source: [76])

Material	Mean annual load (t year^{-1})		Mean annual floodplain deposition flux (t year^{-1})		Mean annual conveyance loss to floodplain storage (%)	
	River Swale	River Aire	River Swale	River Aire	River Swale	River Aire
Suspended sediment	45,158	18,462	16,894	8604	27	32
Cr	1.17	2.51	0.33	0.25	22	9
Cu	3.66	2.76	0.86	0.38	19	12
Pb	29.40	3.66	24.49	1.30	45	26
Zn	32.51	9.99	17.50	2.43	35	20
Total-P	62.54	120.21	9.83	11.48	14	9

Edwards *et al.* [80] demonstrated that remobilization of Cd, Cu, Pb and Zn within overbank sediments of the River Tyne, England, occurred as a result of changes in water-table levels and the breakdown of organic matter above the water table.

Contaminants stored on floodplains also may be remobilized through physical erosion, as shown by Brunet and Astin [81], who reported that elevated discharges of inorganic N and P were associated with increased autumn rainfall and sediment conveyance in the River Adour, southwest France. Physical remobilization may take place long after the primary contaminating activity (e.g. mining) has ceased [41, 82]. Macklin [41] showed that the primary source of Pb and Zn to the contemporary River Tyne, England, was remobilized floodplain alluvium originally deposited during 18th and 19th century metal mining. Macklin [42] stressed that the long-term stability of contaminant metals with respect to changes in physical (river bank and bed erosion, land drainage and development) and chemical conditions (redox and pH) is poorly understood (see also Section 5.2 for effects of climate change on floodplain sediment–contaminant remobilization). There is, therefore, a clear need to quantify these diffuse inputs of historical contaminants in river basin management plans.

Similarly, there can be considerable storage of sediment-associated contaminants within river channels [61, 62, 83], such as on, and in, the channel bed and in 'dead-zones' (also see Section 2.6). Thus, Kronvang *et al.* [84]

document the presence of 19 pesticides (old and modern) and nine heavy metals in the channel bed-sediments of 30 lowland streams in Denmark.

Lakes and reservoirs also represent areas of temporary storage within the sediment–contaminant cascade from source to sink ([85, 86]; Chapter 1, this book). However, they tend to represent sites of significant net storage, and are generally not significant in terms of sediment-associated contaminant *sources* to downstream reaches, although they should be evaluated within any catchment monitoring and/or management programme.

Clearly, the physical remobilization and resuspension of temporarily stored contaminated sediment within the channel system, including floodplains, riparian wetlands, lakes, reservoirs and in-channel sources, introduces the contaminated sediment back into the water column for subsequent redistribution within the river system. Such remobilization can be brought about by changes in hydrological conditions (such as large flood events), changes in hydraulic gradient (due to changes in base/sea level) and disturbance (dredging, construction) (see Section 5). Furthermore, some sediment-associated contaminants (such as P) may be released from sediments to the overlying water column under certain circumstances [87–90], thus representing important sources of contaminants, and this is discussed in more detail below.

3.8. Diagenetic remobilization

Chemical diagenesis in aquatic sediments can lead to the release of contaminant species from sediments, via a series of bacterially mediated, organic matter oxidation reactions. As a result, aquatic sediments may act as a diffuse source of contaminants via this mechanism. These bacterial reactions are the primary control on the release of chemical species and gases (e.g. H_2S, CH_4, NH_4^+, HCO_3^-, Fe^{2+}) from sediments into porewaters. Contaminants associated with these species (e.g. metals adsorbed onto iron oxide surfaces) will also be released by these reactions. Studies in lakes and shallow marine systems have clearly documented the early diagenetic release of contaminants from sediments through these reactions [91, 92]. In many of these studies, the resulting flux of contaminants (termed a 'benthic flux') out of the sediments into overlying waters has been estimated to be of the same order of magnitude as runoff input. For example, fluxes of Co and Zn from sediments in the San Francisco Bay, USA, are of the same magnitude as riverine inputs [92]. Similarly, fluxes of Cd and Zn from coastal sediments in Massachusetts, USA, are a similar magnitude to that within the water column itself [93].

4. Sediment–contaminant transport and transfer processes in rivers

In addition to the need to understand and quantify sediment–contaminant sources and transfers *to* river channels, sediment management at the river basin scale also requires an understanding of the transfer pathways and processes that control sediment and associated contaminant fluxes *through* river channel systems. In particular, information about sediment–contaminant fluxes in rivers is very useful in order to evaluate erosion rates, to describe sediment dynamics at different time scales, to assess the geomorphological and societal implications of river loads (i.e. identifying if there is a problem), and for identifying management needs and developing management plans (i.e. managing the problem). Some in-channel sediment–contaminant sources and redistribution processes associated with the channel bed are discussed in Sections 2.6 and 3.7, and others are considered below.

4.1. Sediment transport processes and fluxes

4.1.1. Continuity of sediment transport in river systems

Viewed over the long term, erosion processes remove sediment from the land surface, and the river network carries the erosional products from drainage basins, ultimately exporting it to the global ocean. Conceptually, the river basin or catchment can be divided into three zones: (a) the erosion or sediment production zone (steep, rapidly eroding headwaters); (b) the transport zone (through which sediment is moved more or less without net gain or loss); and (c) the deposition zone [94] (also see Figure 4, Chapter 1, this book). Within the transport zone, the river channel can be seen as a conveyor belt [95], which transports the erosional products downstream to the depositional sites below sea level. Changes in sediment grain size occur as the sediment moves through the sediment cascade, reflecting hydraulic sorting by water and the effects of weathering and abrasion. Transport of sediment through the catchment and along the river system can be continuous or episodic.

A river channel is a dynamic feature that, together with its floodplain, constitutes a single hydrologic and geomorphic unit characterized by frequent transfers of water and sediment. The failure to appreciate the integral connection of river reaches and storage elements (such as floodplains and reservoirs), via fluxes of water, sediment and chemicals (including contaminants and nutrients), from the headwaters to the deposition zones, often underlies many environmental problems in river management today [96] and is one of the main reasons that sediment management requires a basin scale approach (see Chapters 1 and 2, this book).

The total load of a river can be broadly divided into the particulate (or sediment) load and the dissolved (or solute) load. The latter is usually defined as that material <0.45 μm. However, this simple separation is complicated by the existence of colloids. These are particles whose size is in the range 0.001–1 μm [97], and therefore colloids overlap the dissolved and particulate boundary. The separation between particulate/sediment and dissolved/solute load at 0.45 μm is, however, convenient and is commonly used. The sediment load can be further subdivided into the bedload and suspended load. Whether material is bedload or suspended load is determined by the relationship between flow conditions and the transport process, structure, density and size of the material, with the suspended load being composed of finer and/or less dense material. For further information on sediment composition and size see Chapter 1 (this book).

4.1.2. Transport of coarse sediment as bedload

When the flow conditions exceed the threshold of motion then sediment particles along a river channel bed will start to move. Bedload is an encompassing term for the processes of grain sliding, rolling and saltation, and its movement is a relatively poorly understood phenomenon [98]. Information about bedload transport is very useful for evaluating erosion rates, for describing sediment dynamics, and for assessing the geomorphic implications of sediment transport. This information can also be very valuable for civil engineering purposes. The bedload concept is generally applied to particles larger than 0.1 mm, which move in contact with the channel bed. Bedload moves episodically through the channel and its rate of transport is primarily controlled by stream discharge. The entrainment of channel-bed material is a fundamental process that influences a wide variety of fluvial research problems, such as palaeohydraulic reconstructions, canal design, flushing flows or assessment of aquatic habitat [99, 100]. Most investigators use a standard or modified form of the critical Shields' parameter to define incipient motion of a grain size of interest [101]. The Shields' parameter, or dimensionless critical shear stress (τ_{ci}), is defined as the ratio of the fluid forces tending to initiate motion of a given particle to the gravitational force tending to keep the particle at rest:

$$\tau_{ci} = \tau / (\rho_s - \rho_w)\, g\, D_i \qquad (1)$$

where τ, ρ_s, ρ_w, g and D_i denote bed shear stress (N m^{-2}), sediment density and water density (kg m^{-3}), gravitational acceleration (m s^{-2}) and the percentile diameter of the particle (i), respectively. Shields [102] demonstrated that the τ_{c50} of near-uniform grains varies with critical boundary Reynolds number and hypothesized that it attains a constant value of about 0.056 for rough turbulent flow.

The main problem with the Shields' parameter or curve arises with channel bed sediment deposits of mixed sizes [103–105]. Studies of particle entrainment from mixed-size natural deposits [104–106] have also demonstrated systematic deviations from the standard values initially reported by Shields [102]. The influence of the grain size distribution of bed material upon the critical dimensionless shear stress for entrainment was isolated most notably by field studies conducted during the 1980s [107–110]. They suggested that τ_{ci} is not constant but depends strongly on the ratio of the particle size being entrained to the median size of bed material.

Bed load constitutes an important component of the total sediment yield of a drainage basin, typically ranging between 0 and 50% for most rivers, depending upon whether the local environment is humid or arid and whether the channel bed is composed of sand or gravel. It may have significance beyond its relative contribution in that it is channel-forming, but its denudational role is generally inferior to that of suspended sediment [111]. In sand-bed channels, bedload may constitute a small proportion (1–20%) of the total load [112]. Similar values are reported for gravel-bed rivers [113, 114].

At present, bedload measurements are commonly not available in many rivers, even under steady low flows, and a limited number of attempts have been made to determine bedload discharge and bed-material load. One of the main problems arising from the measurement of bedload transport is that, under natural conditions, bedload discharge is not a steady process and variations up to more than 50% can be expected [115]. Unsteady bedload transport under quasi-steady flow conditions occurs commonly in gravel or mixed sand and gravel-bed streams [116, 117]. All observations that have been made with sufficient temporal resolution reveal short-term fluctuations in transport [118, 119]. Within a stream cross-section there are zones that exhibit higher rates of transport than others and, clearly, this presents a challenge in terms of sampling adequately across a channel width. The temporal variations in bedload transport rates can occur on several scales [111, 116]. Sediment pulses can occur over time periods ranging from seconds up to several months [120]. This non-uniform variation in transport volumes has caused much difficulty in establishing representative sampling procedures [121]. Information on bedload sampling methods for determining fluxes is contained in Parker *et al.* [122].

4.1.3. Transport of fine sediment as suspended sediment

In river channels most of the fine sediment is transported as suspended sediment, which is usually <2 mm (i.e. sand-sized material or less) in size, with much of this being <63 μm (i.e. silt- and clay-sized material) (also see Chapter 1, this book). Walling and Moorehead [123] reviewed suspended sediment particle size information for selected rivers of the world and demonstrated that there is considerable spatial variation in values of median diameter (d_{50}), ranging from <1 μm to >100 μm. As an example, Table 4 presents some information on the particle size composition of the suspended sediment transported by several rivers in England and Scotland. For these rivers, which range in size between 34 km^2 and >8000 km^2, the d_{50} ranges between ca. 4 μm and 14 μm, and the % <63 μm ranges between 91% and 100%. Thus, the values presented in Table 4 suggest that the suspended sediment transported in most rivers in England and Scotland is relatively fine by world standards.

The suspended load is composed of mineral and organic particles. Much of the former is derived from terrestrial sources (i.e. those listed in Section 2), whereas a large component of the organic material, such as periphyton and plankton, is derived from within the river channel. Although, most studies have tended to consider suspended sediment as being composed and transported as individual grains, a considerable body of work is demonstrating that in freshwater systems much of the organic and inorganic suspended load is transported as composite particles or 'flocs' [125–127] (also see Figure 1, Chapter 1, this book). This has important implications for sediment–contaminant fluxes in rivers systems due to differences between composite particles and primary particle grains. Composite particles are different in effective size and density (and therefore settling velocity), and the number of sorption sites available for contaminants, compared to individual grains. The existence and implication of composite particles has long been recognized in estuarine and marine environments, particularly in terms of contaminant and nutrient fluxes [128]. Similarly, recent work is also addressing the role of colloids in sediment and contaminant fluxes in river basins [129, 130], particularly because of their small particle size and large specific surface area.

There is a broad relationship between suspended sediment concentration (and fluxes) and water discharge, with sediment concentrations tending to increase with discharge (also see Section 5.1). This reflects the ability of increases in flow to entrain and transport more and larger particles (see also Section 4.1.2) and the fact that increased river flow is usually commensurate with rainfall or snowmelt, which increases sediment supply to rivers due to soil erosion, etc. The considerable scatter between suspended sediment concentration and discharge is due to a range of factors, including variations in catchment geology, rainfall intensity, and sediment supply from the sources. Within individual floods there can be hysteresis effects due to the exhaustion of the amount of sediment available for transport. Thus, typically, suspended sediment concentrations on the rising limb of a storm hydrograph (i.e. before the peak) are greater than values on the falling limb for an equivalent discharge as sediment sources become exhausted.

Table 4. Particle size information for several rivers in England and Scotland. Values are averages based on numerous samples at each site (source: adapted from [124])

Basin	River	Basin area (km^2)	d_{50} (μm)	%<63 μm	%<2 μm
Humber	Cover	83	9.20	95.4	16.5
	Bishopdale Beck	69	9.98	94.1	15.0
	Bedale Beck	112	10.49	93.9	14.1
	Wiske	215	4.33	98.4	31.8
	Cod Beck	130	7.74	95.0	20.6
	Burn	92	13.54	91.4	10.8
	Thornton	34	5.85	98.3	24.0
	Laver	78	11.31	92.7	12.8
	Upper Swale	499	8.45	95.3	17.9
	Lower Swale	1456	7.70	96.3	19.7
	Ure	926	8.71	96.1	16.8
	Nidd	525	9.18	94.5	16.0
	Ouse	3520	7.98	96.2	18.5
	Wharfe	814	8.70	96.2	17.8
	Aire	1932	6.61	97.6	21.4
	Calder	899	6.44	97.7	22.8
	Don	1256	4.06	99.6	26.7
	Trent	8231	8.96	92.9	15.5
Tweed	Teviot	1500	6.79	98.0	22.1
	Middle Tweed	1100	8.90	96.3	17.3
	Lower Tweed	4390	7.64	97.5	19.8

Similarly, there are temporal variations in the particle size characteristics of the suspended sediment transported in rivers due to changes in the discharge and the supply of sediment of various types during storm events and over longer time periods. Generally, coarser material is transported during higher discharges, although there are many documented cases where the opposite is true or there is no relationship between discharge and suspended sediment particle size during storm events [123, 124], which is thought to reflect the fact that suspended sediment transport in these situations is supply-limited rather than a function of transport capacity or flow hydraulics. Thus, the particle size composition of the suspended sediment load in a river at a point in time is strongly influenced by the particle size composition of the source material, which varies according to type (e.g. topsoil, channel bank material, resuspended bed material). There are also considerable variations in the organic matter (and nutrient and contaminant) content of suspended sediment reflecting variations in the type of material being supplied to the river (including seasonal effects) and river flow characteristics. Although there are strong temporal and spatial variations in the concentration and particle size composition of suspended sediment, at a particular river cross-section studies have shown that concentrations and size composition are generally uniform [124, 131, 132], reflecting the fine-grained nature and low settling velocities of the suspended sediment, and the turbulence and mixing observed across a stream at a point in time.

Estimation of suspended sediment transport and average sediment concentrations requires the integration of continuous data on streamflow with measurement of sediment concentration, and further information on this is contained in Chapter 5 (this book). Available methods for measuring suspended sediment concentrations and fluxes in rivers can be divided into those based upon the collection of suspended sediment samples and those based upon turbidity monitoring, and further information is contained in Parker *et al.* [122].

4.2. Contaminant transport processes and fluxes

4.2.1. Particulate vs dissolved load

Information on the contaminated particle fraction and the interaction between the dissolved and particulate phase of micropollutants is necessary for determining the transport, fate, bioavailability and toxicity of such substances in river ecosystems. A large proportion of the contaminant and nutrient load in river basins is transported by particulate matter. For example, Gibbs [133] suggested that up to 90% of the metal load is transported by sediments in many rivers, but this varies from metal to metal. The ‘dissolved’ portion encompasses contaminants that are truly dissolved or colloids. The ‘particulate’ portion of the contaminant load comprises contaminant-rich grains (e.g. metal sulphide grains from tailings effluent) or contaminant element-bearing Fe and Mn oxide coatings on other particles.

The partitioning of contaminants between the dissolved and particulate load depends on both physical and chemical factors. The chemical factors include sediment mineralogy, adsorption onto fine-grained material, co-precipitation with or sorption on hydrous Fe-Mn oxides and carbonates, association with organic matter, salinity, pH and redox processes. Physical factors influencing contaminant transport and the partitioning between the dissolved and particulate load include sediment shape, site geomorphology, streamflow, suspended sediment concentration, climate and season, water depth, flow dynamics and sediment grain size. Gibbs [133] suggested that grain size is possibly the most significant factor controlling the concentration and retention of contaminants in both suspended and bottom sediment. Metals in particular have been shown to be enriched in the fine silt and clay fractions of sediments as a result of their large surface area, organic and clay contents, surface charge and cation exchange capacity.

4.2.2. Sediment–contaminant dispersion

Once sediment-bound contaminants enter river basins, they are dispersed and transported downstream. Work on mining-contaminated rivers has demonstrated that sediment contaminant metal concentrations tend to decrease downstream from contamination sources in a systematic way [134]. Variations in downstream contaminant concentration can result from floodplain storage or inputs of contaminants from diffuse or other point sources [42]. The downstream decrease in concentration may be ascribed to a variety of factors, including:

- dilution of contaminated sediment by uncontaminated sediment derived from tributaries and erosion of channel banks;
- hydraulic sorting of channel bed sediment on the basis of density, size or shape;
- abrasion of contaminated sediment grains;
- storage of contaminated particles in channel and floodplain deposits; and
- chemical sorption or dissolution of contaminants and/or contaminant uptake by biota.

4.2.3. Influence of tributaries

Tributaries have an influence on the contaminant concentration in the main river (C_{main}) and also on the total load of contaminants if there is an additional input to the main river. As a first approach, this effect can be described by applying a simple mixing equation [135]:

$$C_{main} * Q_{main} + C_{trib} * Q_{trib} = C_{mix} * (Q_{main} + Q_{trib}) \quad (2)$$

with

$$C_{main} = C_{particulate}\ (\text{mg kg}^{-1}) * C_{total\ suspended\ sediment}\ (\text{mg l}^{-1}) \text{ and } Q\ (\text{m}^3\ \text{s}^{-1}) \quad (3)$$

This assumption can be applied if the particulate contaminants are restricted to the grain size fraction smaller than 20 μm and the total suspended sediment grain size distribution of both the rivers, that is the tributary and the receiving river, is unchanged. The mixing approach assumes a complete mixing of the contaminated fraction at the point of confluence in the main river and the joining tributary. In the case of measurements taken in a contaminated plume with lateral concentration gradients, the concentration profile must be known to allow for the evaluation for the mass balance by integration of the product of local concentration and flow velocity. If the mixing approach is acceptable, it can be applied to each point of confluence along a river. By knowing the residence time and specifying some reaction and degradation processes, the final concentration at the end of the river can be estimated.

5. Perturbations to sediment and contaminant sources and transfers

There are numerous natural and anthropogenic influences that cause pronounced changes in the type of sources contributing sediment and contaminants to a river and the rate at which sediment transfers occur within river basins. The following sections list a selection of relevant natural (Sections 5.1 and 5.2) and anthropogenic (Section 5.3) influences that operate at a range of spatial and temporal scales, with particular emphasis on anthropogenic influences.

5.1. Natural short-term events

Natural, short-term (e.g. minutes, hours, days) changes in climatological and hydrological conditions can have a major impact upon the transfers and fluxes of sediment and contaminants. One of the most marked of these is changes in discharge as a result of rainfall events. During such events, increased surface runoff can result in orders of magnitude increases in suspended sediment transfers through the basin. For example, in the Bradford Beck, a tributary of the River Aire in England, Old *et al.* [136] documented the sediment dynamics for a single, large convectional summer rainfall event. During this event, over a period of 15 minutes, water discharge increased from 0.45 to 34.6 $m^3 s^{-1}$. During the same period suspended sediment concentrations reached a peak of 1360 mg l^{-1} and at the peak of this sediment discharge event the sediment flux reached 47 kg s^{-1}. Old *et al.* [136] concluded that although the Bradford Beck catchment represents only 3% of the catchment area of the River Aire, at times it can be the major contributor of fine sediment.

Such flood-related suspended sediment transfers can also have a major impact upon the contaminant flux of a river basin. This is illustrated in Figure 5 for the

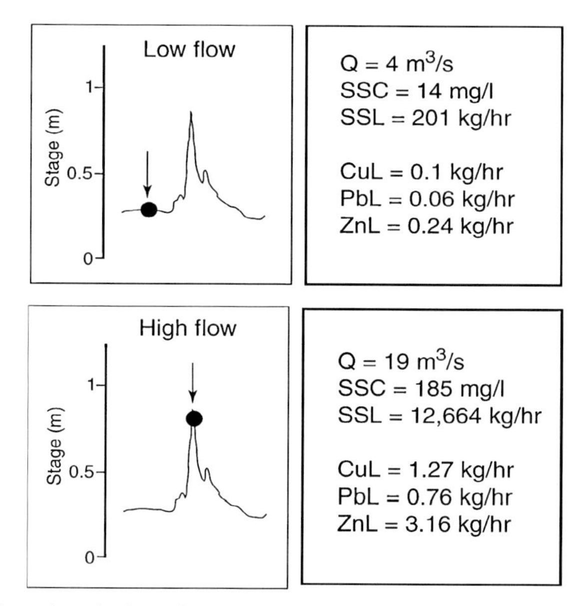

Figure 5. An urban river in low flow and high flow, showing typical stage and suspended sediment relationships (River Medlock, Greater Manchester. Q = discharge, SSC = suspended sediment concentration, SSL = suspended sediment load, CuL = copper load, PbL = lead load, ZnL = zinc load) (source: J. Coyle, published with permission)

River Medlock, a small urban river in Manchester, England. At low flow, suspended sediment concentrations and fluxes are low, with suspended sediment loads of only 20 kg h^{-1}. As a result, these low-flow stages only contribute low levels of contaminant flux. At high-flow events suspended sediment loads of over 12,000 kg h^{-1} can be observed, with resulting high levels of contaminant flux (e.g. over 3 kg h^{-1} of Zn). High-flow events, therefore, have a significant impact on contaminant input into downstream receiving water bodies, such as the lower reaches of rivers, estuaries and the coastal zone.

5.2. Natural long-term events

Climate change involves changes in precipitation and temperature, and the former can have major impacts upon river hydrology and sediment erosion, transport and deposition. In general, increases in precipitation have been correlated to increases in suspended sediment transport [137, 138], but relief, soil, rock and vegetation type play a major role in controlling sediment responses. The response of sediment regime to climate change can be rapid, as has been demonstrated in Holocene and Quaternary fluvial successions [139, 140]. Climatic change in the near future will, therefore, likely have significant impacts upon sediment transfers in river basins, particularly extreme events (see

Section 5.1). For example, Semadeni-Davies [141] modelled the impact of possible future climate change on a cold-climate urban river basin and showed that frequency and volume of waste-water flows in an urban environment would be significantly altered, with implications for drainage system design and management.

Floodplains act as major stores of historical contamination and the re-introduction of contaminants as a result of their erosion has been discussed in Section 3.7. Contaminant remobilization from the erosion of floodplains is often triggered by climate changes [42, 134, 142, 143]. Macklin [42] warned that floodplain contaminant remobilization is increasing as a result of the hydrological changes associated with global warming. Chemical remobilization of contaminants from floodplains can also take place, via changes in groundwater pH and redox [80]. These changes themselves can be linked to changes in climate, as well as tectonic and land-use changes.

5.3. Anthropogenic changes

There are many anthropogenic activities that influence the sources, pathways, storage and fluxes of sediment in river systems. Here we focus on those due to: (a) river engineering, with particular reference to impoundments and reservoirs and in-stream gravel mining; and (b) land-use changes, such as deforestation, agriculture and urbanization, and including the effects of fires. The effects of other types of anthropogenic activities on sediment–contaminant fluxes are discussed in [144–146] and Chapter 1 (this book).

5.3.1. Impoundments and reservoirs

Impoundments, such as dams and associated reservoirs, and artificial lakes disrupt the continuity of a river system, especially its sediment load, by interrupting the conveyor belt of sediment transport from source to sink. Reservoirs and artificial lakes trap all inflowing bedload sediment and a significant part of the suspended sediment load (Figure 6). Sedimentation in reservoirs causes a progressive reduction of dam impoundment capacity, and creates serious problems for water management, especially near dam outlets. The quality of the stored water in the impoundment can also be degraded, due to eutrophication and contaminant fluxes from underlying sediment. Impoundments can also cause downstream changes in river morphology and ecology, the nature of which depends upon the characteristics of the original and altered flow regimes and sediment loads. Dams release sediment-starved or 'hungry' water to downstream reaches, which may transport sand and gravel downstream without replacement from upstream, resulting in coarsening of the surface layer or 'armouring'. Furthermore, because the excess energy may lead

Figure 6. Sedimentation in the Escales Reservoir, Ebro River basin, Spain (photo: R.J. Batalla, February 2004)

to the erosion of the river bed, incision and undercutting of channel banks may result, thereby causing channel widening. The effects of 'hungry water' on downstream river channels can cause dramatic changes on river ecology, such as loss of spawning gravels, and damage to bridges and other infrastructure [96].

In addition, impoundments (especially those designed for flood defence purposes) often diminish the magnitude of downstream floods, which transport the majority of sediment, and reduce the supply of coarse sediment such as sand and gravel to coastlines and deltas. Under such conditions, beaches can become undernourished and shrink, and coastal and delta erosion may be accelerated. In delta regions, for example, the equilibrium between fluvial and marine processes is disrupted, and the delicate balance between sediment deposition and coastal erosion may be changed. The erosion of the Nile Delta in Egypt by about 150 m year^{-1}, 1000 km downstream of the Aswan High Dam, demonstrates the importance of sediment supply from the upper catchment. There are many other similar examples [96]. Beach nourishment with imported sediment dredged from reservoirs and harbours has been implemented along many beaches in southern California, USA, as reported by Kondolf [96]. However, the high costs of transportation, sorting for the relevant particle size fractions, and the cleaning of contaminated dredged materials, as well as the difficulty in securing a stable supply of material, make these options not feasible in many places (also see [147]). With the consideration of fluvial sediment supply and budgets in mind, and the maintenance of coastal beaches in the existing legal framework, a system of 'sand rights', analogous to water rights, has been proposed [148].

Sedimentation in reservoirs in California causes a progressive reduction of dam impoundment capacity, at an estimated US$6 billion in replacement costs every year [149]. In addition, it creates serious problems for reservoir operation, especially near to the outlets of dams. Hydropower facilities suffer the most due to the sediment progressively blocking paths to the power plant. The passage of sediment to the power plant can also erode turbines, requiring frequent repair of plant facilities.

Reservoir sedimentation is a worldwide problem, but is especially problematic in areas with high sediment yields and a high degree of impoundment, such as Mediterranean climatic regions. In such areas, water availability and demand are out-of-phase and thus the degree of river impoundment tends to be higher than in more humid regions. Most dams are built for water storage instead of flood control. For example, Spain owns more dams than any other country in Europe and 2.5% of the total number of dams in the world. Surveys indicate that almost 10% of reservoirs in Spain have experienced a reduction in capacity of 50% or more [150]. Assuming that 50% of the remaining reservoirs have no significant sedimentation problems, the mean weighted reduction of reservoir capacity in Spain can be estimated around 10%. Dams and reservoirs built during the 20th century in Spain are, on average, 35 years old. Combining these two figures gives a mean annual reduction of reservoir capacity of 0.3%, with sedimentation of 170×10^6 m^3 $year^{-1}$. Taking the Ebro River basin as an example, using the same sedimentation rates as for the whole of Spain, but taking into account that reservoirs in the Ebro basin are on average older (50 years), the mean annual reduction of reservoir capacity would be 0.2%, giving an estimated total annual sedimentation of 15×10^6 m^3 [151]. The same sedimentation rate (0.2%) can be obtained from data reported by Sanz *et al.* [152] for 17 reservoirs experiencing sedimentation problems and representing 50% of basin impoundment capacity. Sediment retained in reservoirs along the Ebro basin is composed mainly of silt (62%), clay (25%) and sand (13%).

Some reservoirs in the Ebro basin are already full of sediment, such as the Pignatelli reservoir, constructed in 1790 with an original capacity of 1×10^6 m^3 and the Escuriza reservoir in the Martín River, constructed in 1890, with an original capacity of 6×10^6 m^3. In other cases, sedimentation has also been recognized as a problem, such as the Terradets reservoir on the Noguera Pallaressa, constructed in 1935, with an actual capacity of 8×10^6 m^3 from an original capacity of 23×10^6 m^3. At present, the Mequinenza and Riba-roja reservoirs, some of the largest in the basin, capture around 1×10^6 t $year^{-1}$, of which 98% is transported as suspended load and 2% as bedload [153]. This value agrees with calculations on bedload transport capacity reported by Guillen *et al.* [154]. Palanques and Drake [155] reported a total sediment load entering the Mequinenza and Riba-roja systems of 0.5×10^6 t year^{-1}, while Sanz *et al.* [152] calculated 1×10^6 t year^{-1} based on suspended sediment sampling. The lower Ebro River does not receive coarse fractions from the upstream Mequinenza and Riba-roja reservoirs, but partially keeps its bedload transport capacity since floods have not been dramatically reduced. Bedload is entrained from channel bed and lateral deposits, and are deposited at the lowermost reaches in the delta plain. In addition, over recent years, the river-bed has been almost continuously dredged at a rate of 40,000 t year^{-1} to ensure the navigability of the river for tourism, which has resulted in the degradation of fish habitats and has contributed to the disequilibrium of the river's sedimentary system. Altogether, of the order of 200,000 t year^{-1} of sediment is discharged from the Ebro River to the sea. It is estimated, however, that this represents only 1.5% of what was delivered at the beginning of the 20th century. Lack of sediment nourishment due to large dam construction and reduction in sediment transport capacity have been identified as the main reasons for the retreat of the Ebro delta, which has been observed since the 1970s. Overall, sediment deficit downstream of dams has been estimated at 1×10^6 t year^{-1} [153]. Sediment deficit causes riverbed adjustments such as bank erosion, and channel narrowing and incision, the latter at an estimated average rate of 20 mm year^{-1}.

5.3.2. Gravel mining

Instream gravel mining is, together with river impoundments, the main cause for sediment deficit in many rivers and coastlines. Sand and gravel are used for construction purposes and they are derived primarily from alluvial deposits, usually directly from the river bed (see Figure 1). For instance, in Spain fluvial mining has taken place under almost no control, creating a huge sediment imbalance in many rivers, and many of these rivers are also regulated. Instream mining directly alters the channel geometry and bed elevation, while disrupting the continuum of sediment transport downstream. The main effects of gravel mining are *in situ*, but there are also downstream effects, such as [96]:

- channel incision and bed coarsening, in a similar manner to the way that a dam disrupts the pre-existing balance between sediment supply and transport capacity;
- undermining of structures, as a direct effect of river-bed incision and lateral channel instability (Figure 7);

Figure 7. The Carlos III bridge (18th century) in the Llobregat River, Spain, after its collapse during a flood in 1971. Lack of sediment feeding from upstream reaches due to gravel mining was the main cause (source: [151]. The photo is reproduced with kind permission of the Spanish Society of Geomorphology and the Spanish Quaternary Association)

- destruction of fish habitat in the mining areas and downstream reaches, by the modification of pool-riffle distribution and alteration of inter-gravel flow paths; and
- lack of sediment supply to coastline and delta areas: due to the high cost of transportation, many aggregate supply operators concentrate their activities in alluvial zones near the areas of consumption, typically near cities and tourist centres.

Annual consumption of aggregate is estimated at around 7 t per person, the second largest natural resource use after water (see www.aridos.org, www.gremiarids.com), and at a market mean value of 7 € t^{-1}. Damage to bridges and other infrastructure due to gravel mining is not directly incorporated in the price of aggregate. In the case of the River Tordera, Spain, the ratio between cost of bridge reconstruction and reinforcement and the mass of sand and gravel extracted is estimated to be of the order of 2–4 € t^{-1}, which is similar to the US$5 of post-failure public investment per t of gravel in the San Benito River in California (Kondolf, personal communication).

Gravel mining has been particularly intense in some tributaries of the Ebro River, which already exhibit evidence of severe sediment deficit due to upstream regulation. Examples of such extreme mining activity can be seen elsewhere in the main channel of the River Segre and its tributaries, such as the Noguera Pallaresa, the Noguera Ribagorçana and the Ribera Salada, in which the sediment by volume extracted within the last 20 years is 200 times higher than the annual mean bedload yield of ca. 2000 t year^{-1}. The Siurana River, the main downstream tributary of the Ebro River, suffers from intensive gravel mining downstream of the Siurana Dam (Figure 8), especially in its lowermost reach, reducing its sediment contribution to the main channel of the Ebro River to virtually nothing. This further accentuates the sediment deficit of the Lower Ebro River.

One of the most dramatic examples of gravel mining in the Catalan Coastal Ranges, Spain, is the unregulated Tordera River (970 km^2). There, approximately 5×10^6 t of sand and gravel were extracted during the 1960s and 1970s until 1982 when mining was prohibited. This value is more than 10 times the annual sediment yield of the Tordera River, including both suspended and bedload [156]. Fluvial sediments were converted to aggregate for construction in the Costa Brava area during the rapid growth of tourism during those decades. The consequences of intensive gravel mining are numerous and include massive destruction of river ecosystems, channel incision with the consequent undermining of several bridges (some of which collapsed), groundwater overdrafting, and the severe lack of sediment nourishment to the delta and beaches stretching from Blanes to Barcelona.

Figure 8. Bridge undermining due to gravel extraction in the Siurana River, Spain, upstream of the confluence with the Ebro River, NE Spain (photo: R.J. Batalla, June 2005)

5.3.3. Land-use changes including urbanization

Soil erosion rates and sediment fluxes and yields in river basins vary markedly between basins with different land uses but similar lithological and climatic conditions. Probably one of the most pronounced changes in erosion rates is due to deforestation and Table 5 presents examples of the erosional response to deforestation based on lake sediment-based reconstructions for a variety of different locations. Table 5 illustrates that deforestation results in a dramatic increase in rates of soil erosion. This is because the protective vegetation cover is removed and the soil is exposed to water (especially rainfall detachment and overland flow) and wind erosion processes. Deforestation not only increases rates of soil erosion, but removal of the forest cover also increases other sediment production processes, such as mass movements and channel bank erosion, such that the amount of sediment delivered to rivers increases dramatically. In consequence, sediment fluxes in rivers and sediment delivery to the coastal zone also tend to increase markedly. In Australia, for example, McCulloch *et al.* [157] used the Ba/Ca ratio in long-lived coral to determine that there had been a 5- to 10-fold increase in sediment fluxes to the Great Barrier Reef after the beginning of European settlement in 1870 due to land-use practices such as forest clearing and overstocking. For further information on changes in sediment fluxes see Chapter 1 (this book).

In many European countries, deforestation occurred hundreds of years ago, and river systems are now mainly responding to changes in agricultural land use and management, including the trend for intensification and the application of artificial chemicals to increase food and biomass production. As an example, Figure 9 illustrates the reconstructed changes in the sediment yield delivered to a series of lakes and reservoirs in the UK over the last ca. 100 years, which reflects soil erosion and sediment fluxes in the upstream contributing catchments. It is clear from Figure 9 that sediment yields have increased over

Table 5. Examples of the increase in erosion due to deforestation based on lake sediment-based reconstruction (source: modified from [158])

Country	Forcing	Response	Time (years)
Germany	Clearance from AD 1050	x 10–17	ca. 250
Sweden	Initial clearance from 800 BC	x 4	ca. 200
Vermont, USA	18th century settlement	x 4	ca. 100
Michigan, USA	19th century settlement	x 4	ca. 10
Tanzania	19th century clearance	x 4	ca. 10
Papua New Guinea	19th century clearance and gardening	x 10	>150

time, such that yields in the mid to latter half of the 20th century are greater than those before by a factor of between 2 and 10. Analysis of historical records of land use and climate for these river basins demonstrates that changes in land use and management, especially deforestation and conversion of grassland to arable land, can explain the observed trends in sediment yield, although changes in weather patterns are also important [159]. Foster [159], and Foster and Lees [160] also investigated temporal changes in the fluxes of sediment-associated contaminants and nutrients, such as P, to the same lakes and found that there was evidence for increased fluxes due to land-use and management changes over the same time period.

The effects of urbanization, building activity and road construction on catchment hydrology and sediment fluxes have been widely documented [161, 162]. Batalla and Sala [163] documented downstream within-channel and overbank deposition, and cross-sectional narrowing in small tributaries of the Arbúcies River, Spain, as a consequence of the construction of a new motorway at the headwaters of the basin, just months after the works began. Although several correction steps against erosion were taken during the construction of the road, an average deposition of 20 cm of sediment on the channel and banks at the Riudecos tributary was measured after several storm events in autumn 1994, due to the upstream clearing, and subsequent removal and fill-up of sediment, estimated at more than 3×10^6 m^3 for the entire area.

Urbanization, building activity and road construction activities are also particularly worthy of mention because the sediment supplied to rivers is often contaminated. Owens and Walling [55], for example, used floodplain sediment cores to reconstruct the metal (including Cr, Cu, Pb, Sr and Zn) and P content of the suspended sediment being transported by rivers in the Aire basin, England. They found that temporal patterns of the metal and P content of the sediment over the last 100–150 years reflected the influence of urbanization and the type of industry in the catchment over this period. In most studies, however, once the construction and building phases have been completed, sediment production and fluxes would be expected to decline to levels similar to those of the pre-building period. For many contaminants, the end of the building and construction phase may not result in a concomitant decrease in concentrations, and indeed values may even increase due to changes or intensification of industrial activities and the introduction of other point sources such as STWs.

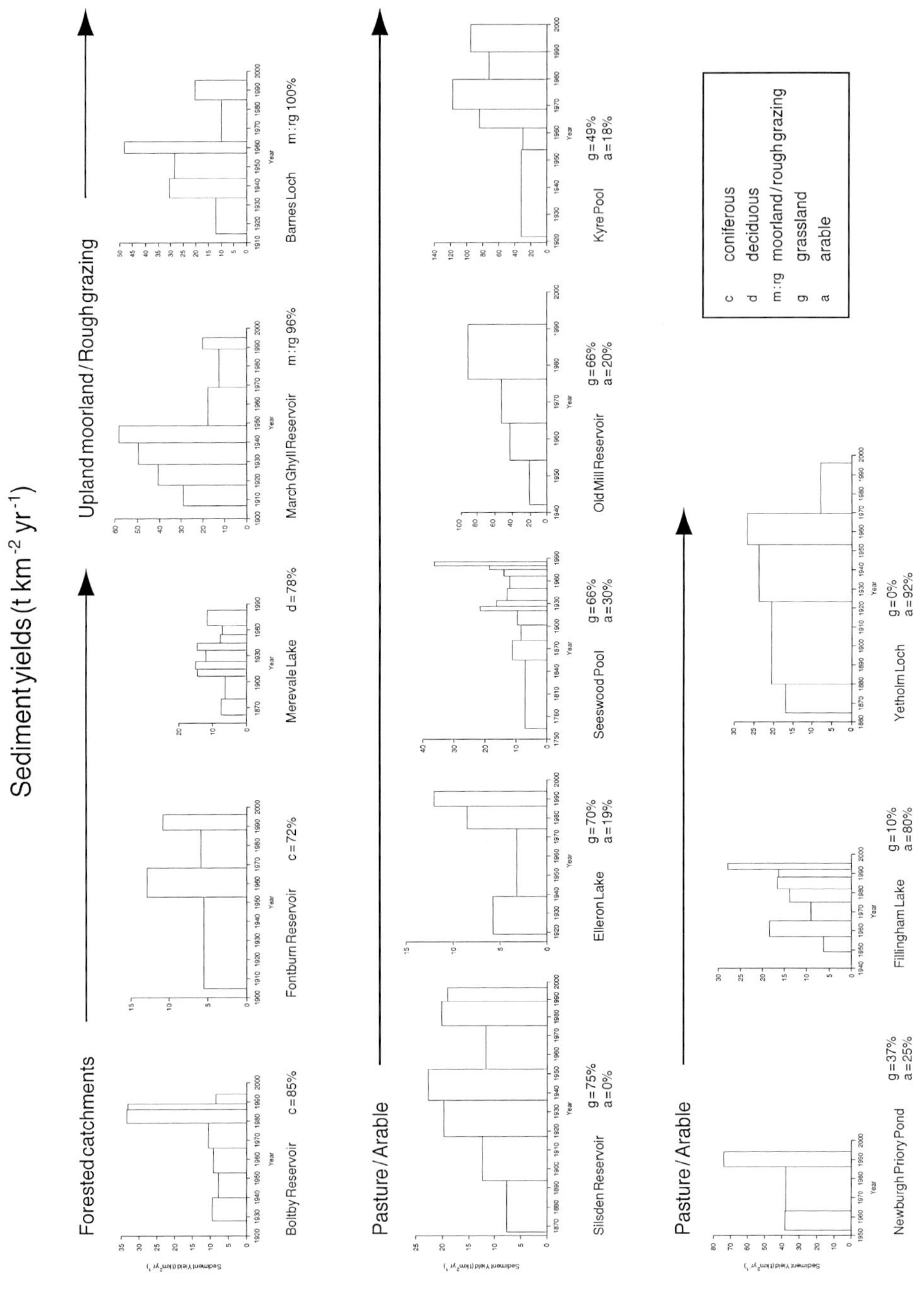

Sediment yields ($t\ km^{-2}\ yr^{-1}$)

Forested catchments

Upland moorland / Rough grazing

Boltby Reservoir c = 85%

Fontburn Reservoir c = 72%

Merevale Lake d = 78%

March Ghyll Reservoir m : rg 96%

Barnes Loch m : rg 100%

Pasture / Arable

Silsden Reservoir g = 75% a = 0%

Elleron Lake g = 70% a = 19%

Seeswood Pool g = 66% a = 30%

Old Mill Reservoir g = 66% a = 20%

Kyre Pool g = 49% a = 18%

Pasture / Arable

Newburgh Priory Pond g = 37% a = 25%

Fillingham Lake g = 10% a = 80%

Yetholm Loch g = 0% a = 92%

c coniferous
d deciduous
m : rg moorland / rough grazing
g grassland
a arable

Figure 9 (opposite). Reconstructed fine-grained sediment yields for several lakes and reservoirs with differing land uses in the UK. In most cases sediment yields increase with time and are linked primarily to changes in land use and management (source: [159] reproduced with the permission of CABI)

5.3.4. Fires

Fires can be broadly classified into two main types: (a) prescribed fires, which are controlled and are usually part of a forest management programme; and (b) wildfires, which are not normally planned and can be either natural (i.e. due to lightning) or due to human activity (i.e. accidental). Either type of fire can cause an increase in soil erosion rates and sediment fluxes in rivers, due to a number of reasons, including:

- Burning-off of leaves and the undergrowth exposes the soil, thus allowing raindrops to strike the soil with a much greater impact than previously.
- Destruction of forest litter and other debris on the ground permits higher overland flow velocities, which in turn permits the entrainment of a greater quantity of sediment.
- Changes in the properties of soils (e.g. soil hydrophobicity), which in turn affects soil hydrological pathways.
- Increase in the amount of exposed topsoil and subsoil subject to erosion processes.
- Increase in mass movement events.
- Increase in channel discharges and flow velocities and associated increase in channel bed and bank erosion.

Fires affect catchment sediment dynamics in two main ways: (a) by changing the main sediment sources; and (b) by increasing sediment fluxes. There are numerous studies that have reported the downstream effects on sediment fluxes at the plot, hillslope and sub-catchment scales, although the effects of fires on soil erosion and sediment flux processes is still relatively poorly understood at the catchment scale. Table 6 lists the effects of fires on sediment fluxes for various river basins, some of which are discussed below. Information on changes in the contributions from the dominant sediment sources due to fires is less readily available, although Rice (cited in [164]) identified the direct sources of sediment produced from Harrow Canyon, California, USA, in 1969. Rice estimated that 74% of the sediment came directly from scour of residual sediment in the Harrow Canyon channel, another 22% came from rills and gullies, and very small quantities came from wind, dry ravel (the rolling, bouncing and sliding of individual particles down a slope; an important process on hillslopes in steep arid and semi-arid environments) and landslides. He identified that major sources of sediment produced from the Harrow Canyon channel originated upslope, and estimated that landslides originally contributed 54% and dry ravel contributed 33% of the sediment. Blake *et al.* [165, 166] used mineral magnetic measurements to investigate sediment sources after wildfires in Australia, and found ephemeral gullies supplying material from burnt ridge tops were particularly important. Similarly, Owens *et al.* [167] have attempted to investigate changes in sediment sources due to a wildfire in British Columbia, Canada, using the sediment fingerprinting approach (for further details of this approach see Chapter 5, this book).

Information on river turbidity responses to fire is also relatively limited. Turbidity is a difficult parameter to characterize because it is highly transient and extremely variable [164]. Most studies to date have reported significant increases in sediment fluxes following fires (Table 6), although there are exceptions (e.g. [174]) due to the lack of rainfall and snowmelt events (i.e. the driving events) in the first few months following the fire. Brown [173] reported increases in stream sediment concentrations between one and three orders of magnitude after a wildfire in New South Wales, Australia. The recovery to pre-fire values was relatively slow and it was not until four years later that observations again plotted close to initial concentrations. Megahan *et al.* [169] reported an increase in sediment yield of 66 times the long-term average following prescribed burning for otherwise undisturbed watersheds in Idaho, USA. Average annual sediment yields showed a statistically significant increase following logging and burning of 7 t km^{-2} or 100%. Accelerated sedimentation showed no signs of abating 10 years after disturbance. Sediment yields were not the result of channel erosion caused by increased streamflow, but by active mass movement events and surface erosion. The magnitude and the duration of accelerated sedimentation caused concern for both downstream fish populations and onsite biomass productivity.

Batalla [172] analysed the effects of the 1994 wildfire on sediment fluxes in the granitic, Mediterranean, Arbúcies basin in Spain. In this basin, approximately 1000 ha (10% of the catchment area) were completely burnt. Results indicated that:

- sediment availability increased dramatically upstream of the basin outlet due to fire, as indicated by the less hydraulic-dependent relation between sediment concentration and discharge (Figure 10);

Table 6. Summary of effects of fire on increases in suspended sediment load (source: modified from [164])

Location	Scale	Vegetation	Increment (%)	Reference
Western Montana, USA	Plot	Western larch, Douglas-fir	16,800	[164][a]
Central Arizona, USA		Chaparral	116,500	[164][a]
Texas, USA	Catchment	Oak-juniper	140,000	[164][a]
California, USA		Ponderosa Pine	0	[164][a]
North Carolina, USA	Plot	Southern woodland	1,120,000	[164][a]
Mississippi, USA	Plot	Oak-woodland	1,650	[164][a]
California, USA	Plot	Chaparral	1,000	[164][a]
Northern Arizona, USA		Ponderosa Pine	41,800	[164][a]
SW Cape, South Africa	Catchment	Mountain Fynbos	0	[168]
Idaho, USA	Catchment	Douglas-fir	97	[169]
Mount Carmel, Israel	Plot	Sclerophyll forest	100,000	[170]
Prades mountains, Spain	Plot	Evergreen Oak	950	[171]
Montseny massif, Spain	Catchment	Evergreen Oak	600–2,300	[172]
NSW, Australia	Catchment	Sclerophyll forest	100,000	[173]

[a]References contained within [164].

- suspended sediment concentrations during the flood immediately after the fire rose by an average of two orders of magnitude compared with characteristic pre-fire values, increasing from 22 mg l^{-1} to more than 2000 mg l^{-1} for a mean discharge of 800 l s^{-1}; and
- suspended sediment loads at peak flows were up to 1000% greater in relation to those obtained for similar discharges prior to the fire.

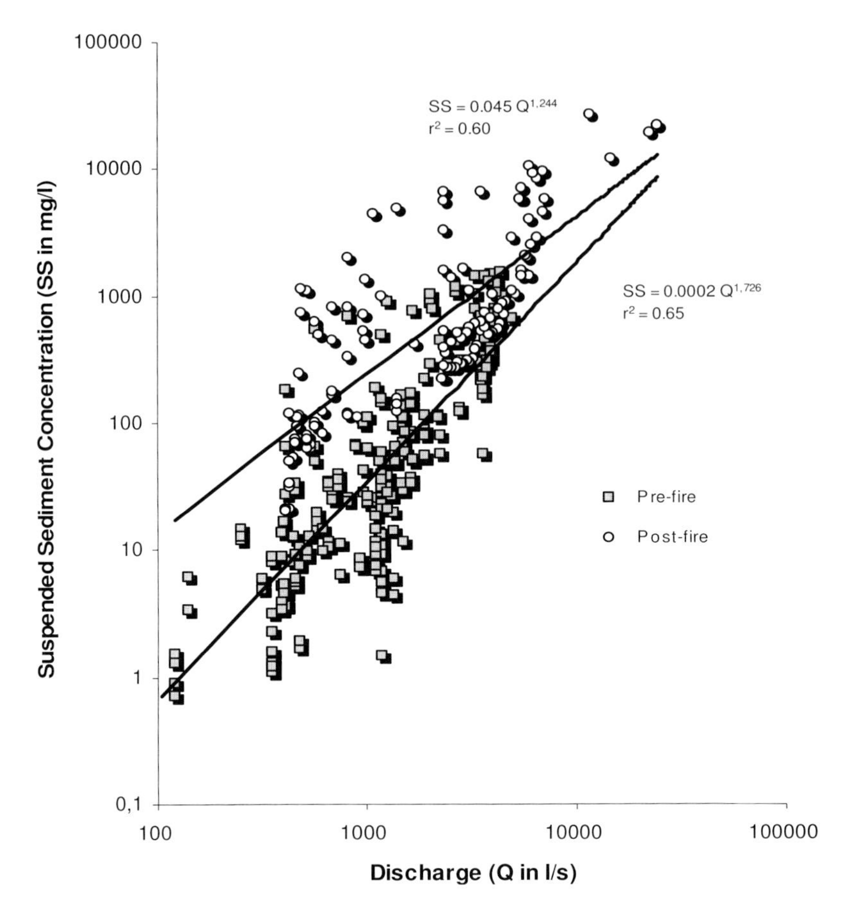

Figure 10. Changes in suspended sediment concentrations after the 1994 wildfire in the Arbúcies basin, Spain (source: [172] reproduced with permission of Universitat de Girona)

The maximum measured suspended sediment concentration during the study period was 27,000 mg l^{-1}, for a discharge of 11.5 m^3 s^{-1} during the recession of the flood event on 12 October 1994, a value that represents more than 15 times the pre-fire concentration for the same discharge. Sediment losses were particularly high in several burnt subcatchments, where quick runoff response caused severe erosion compared to neighbouring forested areas. Sediment contributions from those areas were 16 kg ha^{-1} (from those with a burnt area of 65%) and 65 kg ha^{-1} (from those with burnt area of 100%); these values are 1.5 to 6 times higher than those obtained simultaneously at the downstream basin outlet. Fire, therefore, produced acute effects on the sediment fluxes in the Arbúcies River, taking into account that only 10% of the area was seriously affected. Two facts appear to be responsible for the rapid and strong impact of fire on river sediment fluxes in this situation: (a) the proximity of burnt areas to the sampling point at the basin outlet; and (b) the occurrence of six flood events, the largest with a recurrence period of 8 years, during the weeks immediately following the forest fire.

6. Conclusion

This chapter has briefly reviewed most of the main sources and pathways of both sediment and contaminants in river systems, has described sediment–contaminant transport processes in rivers, and has assessed some of the main natural and anthropogenic perturbations to these sources, pathways and transport processes. This information enables us to understand the sediment–contaminant system so as to effectively target management and to mitigate the detrimental impacts (whether real or perceived) of sediment–contaminant dynamics in river basins. Furthermore, it allows us to assess the responses of the sediment–contaminant system to any interventions, whether they be site-specific or basin scale. Such information, therefore, represents an important requirement for sediment management frameworks and plans (see Chapters 2 and 8, this book). The following chapter (Chapter 5, this book) describes some of the tools and approaches that are available for assembling this information. The task of obtaining a complete understanding of sediment–contaminant sources, pathways and transport processes at the river basin scale probably represents one of the greatest challenges facing those concerned with sediment management. Without a comprehensive understanding of the system that we are trying to manage, it is unlikely that we will ever have the knowledge to make the best decisions. For further information on sediment–contaminant sources, pathways and transport processes, the reader is directed to Garcia and Batalla [175], Owens and Collins [176], Perry and Taylor [177] and Van der Perk [178].

References

1. Jones RJA, Le Bissonnais Y, Bazzoffi P, Díaz JS, Düwel O, Loj G, Øygarden L, Prasuhn V, Rydell B, Strauss P, Üveges JB, Vandekerckhove L, Yordanov Y (2004). Nature and extent of soil erosion in Europe. In: Van-Camp L, Bujarrabal B, Gentile A-R, Jones RJA, Montanarella L, Olazabal C, Selvaradjou S-K (Eds), Reports of the Technical Working Groups Established Under the Thematic Strategy for Soil Protection. EUR 21319 EN/1, Office for the Official Publications of the European Commission, Luxembourg, pp. 145–185.
2. Nihlen T, Mattsson JO (1989). Studies of eolian dust in Greece. Geografiska Annaler, 71A: 269–274.
3. Owens PN, Slaymaker O (1997). Contemporary and post-glacial rates of aeolian deposition in the Coast Mountains of British Columbia, Canada. Geografiska Annaler, 79A: 267–276.

4. Pye K (1987). Aeolian Dusts and Dust Deposits. Academic Press, London.
5. Gerrard AJ (1990). Mountain Environments: An Examination of the Physical Geography of Mountains. MIT Press, Cambridge, MA.
6. Owens PN, Slaymaker O (Eds) (2004). Mountain Geomorphology. Arnold, London.
7. Morgan RPC (2005). Soil Erosion and Conservation, 3rd edn. Blackwell, Oxford.
8. He Q, Owens PN (1995). Determination of suspended sediment provenance using caesium-137, unsupported lead-210 and radium-226: a numerical mixing model approach. In: Foster IDL, Gurnell AM, Webb BW (Eds), Sediment and Water Quality in River Catchments. Wiley, Chichester, pp. 207–227.
9. Collins AL, Walling DE, Leeks GJL (1997). Fingerprinting the origin of fluvial suspended sediment in larger river basins: combining assessment of spatial provenance and source type. Geografiska Annaler, 79A: 239–254.
10. Wallbrink PJ, Murray AS, Olley JM (1998). Determining sources and transit times of suspended sediment in the Murrumbidgee River, New South Wales, Australia using fallout ^{137}Cs and ^{210}Pb. Water Resources Research, 34: 879–887.
11. Matisoff G, Bonniwell EC, Whiting PJ (2002). Soil erosion and sediment sources in an Ohio watershed using beryllium-7, cesium-137 and lead-210. Journal of Environmental Quality, 31: 54–61.
12. Walling DE (2005). Tracing suspended sediment sources in catchments and river systems. The Science of the Total Environment, 344: 159–184.
13. Collins AL, Walling DE, Sichingabula HM, Leeks GJL (2001). Suspended sediment sources in a small tropical catchment and some management implications. Applied Geography, 21: 387–412.
14. Walling DE, Owens PN, Leeks GJL (1999). Fingerprinting suspended sediment sources in the catchment of the River Ouse, Yorkshire, UK. Hydrological Processes, 13: 955–975.
15. Quine TA, Walling DE, Chakela QK, Mandiringana OT, Zhang XB (1999). Rates and patterns of tillage and water erosion on terraces and contour strips: evidence from caesium-137 measurements. Catena, 36: 115–142.
16. Van Oost K, Van Muysen W, Govers G, Deckers J, Quine TA (2005). From water to tillage erosion dominated landscape evolution. Geomorphology, 72: 193–203.
17. Gruszowski KE, Foster IDL, Lees JA, Charlesworth SM (2003). Sediment sources and transport pathways in a rural catchment, Herefordshire, UK. Hydrological Processes, 17: 2665–2681.
18. Carter J, Owens PN, Walling DE, Leeks GJL (2003). Fingerprinting suspended sediment sources in a large urban river system. The Science of the Total Environment, 314–316: 513–534.
19. Owens PN, Walling DE, Leeks GJL (1999). Deposition and storage of fine-grained sediment within the main channel system of the River Tweed, Scotland. Earth Surface Processes and Landforms, 24: 1061–1076.
20. Wilson AJ, Walling DE, Leeks GJL (2004). In-channel storage of fine sediment in rivers in southwest England. In: Golosov V, Belyaev V, Walling DE (Eds), Sediment Transfers Through the Fluvial System. IAHS Publication 288. IAHS Press, Wallingford, pp. 291–299.
21. Phillips JM, Walling DE (1999). The particle size characteristics of fine-grained channel deposits in the River Exe basin, Devon, UK. Hydrological Processes, 13: 1–19.
22. Petticrew EL, Biickert SL (1998). Characterization of sediment transport and storage in the upstream portion of the Fraser River (British Columbia, Canada). In: Summer W, Klaghofer E, Zhang W (Eds), Modelling Soil Erosion, Sediment Transport and Closely Related Hydrological Processes. IAHS Publication 249. IAHS Press, Wallingford, pp. 383–391.

23. Petticrew EL, Arocena JM (2003). Organic matter composition of gravel-stored sediments from salmon-bearing streams. Hydrobiologia, 494: 17–24.
24. Townend I, Whitehead P (2003). A preliminary net sediment budget for the Humber Estuary. The Science of the Total Environment, 314–316: 755–767.
25. Brils J, Salomons W (Eds) (2004). Contaminated Sediments in European River Basins. SedNet Report. TNO, The Netherlands.
26. Correll DL (1998). The role of phosphorus in the eutrophication of receiving waters: a review. Journal of Environmental Quality, 27: 261–266.
27. Smol JP (2002). Pollution of Lakes and Rivers: A Paleoenvironmental Perspective. Arnold, London.
28. Barceló D, Petrovic M (Eds) (2007). Sustainable Management of Sediment Resources: Sediment Quality and Impact Assessment of Pollutants. Elsevier, Amsterdam.
29. Salomons W, Förstner U (1984). Metals in the Hydrocycle. Springer, Berlin.
30. Horowitz AJ (1991). A Primer on Sediment Trace-Element Chemistry. Lewis, Michigan.
31. Foster IDL, Charlesworth SM (1996). Metals in the hydrological cycle: trends and explanation. Hydrological Processes, 10: 227–261.
32. Kronvang B (1990). Sediment-associated phosphorus transport from two intensively farmed catchment areas. In: Boardman J, Foster IDL, Dearing JA (Eds), Soil Erosion on Agricultural Land. Wiley, Chichester, pp. 313–330.
33. Novotny V (2003). Water Quality, Diffuse Pollution and Watershed Management, 2nd edn. John Wiley and Sons, New York.
34. Vink RJ, Behrendt H, Salomons W (2000). Present and future quality of sediments in the Rhine catchment area – heavy metals. Project Report. Dredged Material in the Port of Rotterdam – Interface between Rhine Catchment Area and North Sea. GKSS Research Centre, Geesthacht, Germany.
35. Defra (2002). The Government's Strategic Review of Diffuse Water Pollution from Agriculture in England: Agriculture and Water: A Diffuse Pollution Review. Department for the Environment, Food and Rural Affairs, London.
36. Vink, R. (2002). Heavy metal fluxes in the Elbe and Rhine river basins: analysis and modelling. PhD thesis, Vrije University, Amsterdam.
37. De Cort M, Dubois G, Fridman ShD, Germenchuk MG, Izrael YuA, Janssens A, Jones AR, Kelly GN, Kvasnikova EV, Matveenko II, Nazarov IM, Pokumeiko YuM, Sitak VA, Stukin ED, Tabachny LYa, Tsaturov YuS, Avdyushin SI (1998). Atlas of caesium deposition in Europe after the Chernobyl accident. European Commission Report EUR 16733, Luxembourg.
38. Armour-Brown A, Nichol I (1970). Regional geochemical reconnaissance and the location of metallogenic provinces. Economic Geology, 65: 312–330.
39. Levinson AA (1974). Introduction to Exploration Geochemistry. Applied Publishers, Calgary.
40. Rawlings BG, O'Donnell K, Ingham M (2003). Geochemical survey of the Tamar catchment (south-west England). British Geological Survey Report, CR/03/027, Keyworth, Nottingham.
41. Macklin MG (1992). Metal contaminated soils and sediment: a geographical perspective. In: Newson MD (Ed.), Managing the Human Impact on the Natural Environment: Patterns and Processes. Belhaven Press, London, pp. 172–195.
42. Macklin MG (1996). Fluxes and storage of sediment-associated heavy metals in floodplain systems: assessment and river basin management issues at a time of rapid environmental change. In: Anderson MG, Walling DE, Bates PD (Eds), Floodplain Processes. John Wiley and Sons, Chichester, pp. 441–460.

43. Grimalt JO, Ferrer M, Macpherson E (1999). The mine tailings accident in Aznalcollar. The Science of the Total Environment, 242: 3–11.
44. Hudson-Edwards KA, Macklin MG, Jamieson HE, Brewer PA, Coulthard TJ, Howard AJ, Turner J (2003). The impact of tailings dam spills and clean-up operations on sediment and water quality in river systems: The Ríos Agrio-Guadiamar, Aznalcóllar, Spain. Applied Geochemistry, 18: 221–239.
45. Boult S, Collins DN, White KN, Curtis CD (1994). Metal transport in a stream polluted by acid mine drainage; the Afon Goch, Anglesey, UK. Environmental Pollution, 84: 279–284.
46. Hopke PK, Lamb RE, Natusch FS (1980). Multielemental characterization of urban roadway dust. Environmental Science and Technology, 14: 164–172.
47. Fergusson JE, Ryan DE (1984). The elemental composition of street dust from large and small urban areas related to city type, source and particle size. The Science of the Total Environment, 34: 101–116.
48. Wei C, Morrison GM (1994). Platinum in road dusts and urban river sediments. The Science of the Total Environment, 147: 169–174.
49. Taylor KG, Boyd NA, Boult S (2003). Sediments, porewaters and diagenesis in an urban water body, Salford, UK: impacts of remediation. Hydrological Processes, 17: 2049–2061.
50. Boult S, Rebbeck J (1999). The effects of eight years aeration and isolation from polluting discharges on sewage- and metal-contaminated sediments. Hydrological Processes, 13: 531–547.
51. Owens PN, Walling DE (2002). The phosphorus content of fluvial sediment in rural and industrialized river basins. Water Research, 36: 685–701.
52. Bromhead JC, Beckwith P (1994). Environmental dredging on the Birmingham Canals: water quality and sediment treatment. Journal of the Institution of Water and Environmental Management, 8: 350–359.
53. Kelderman P, Drossaert WME, Min Z, Galione LS, Okonkwo LC, Clarisse IA (2000). Pollution assessment of the canal sediments in the city of Delft (the Netherlands). Water Research, 34: 936–944.
54. Nageotte SM, Day JP (1998). Lead concentrations and isotope ratios in street dust determined by electrothermal atomic absorption spectrometry and inductively coupled plasma mass spectrometry. Analyst, 123: 59–62.
55. Owens PN, Walling DE (2003). Temporal changes in the metal and phosphorus content of suspended sediment transported by Yorkshire rivers, UK over the last 100 years, as recorded by overbank floodplain deposits. Hydrobiologia, 494: 185–191.
56. Robertson DJ, Taylor KG, Hoon SR (2003). Geochemical and mineral characterisation of urban sediment particulates, Manchester, UK. Applied Geochemistry, 18: 269–282.
57. Motelica-Heino M, Rauch S, Morrison GM, Donard OFX (2001). Determination of palladium, platinum and rhodium concentrations in urban road sediments by laser ablation-ICP-MS. Analytica Chimica Acta, 436: 233–244.
58. Whiteley JD, Murray F (2003). Anthropogenic platinum group element (Pt, Pd and Rh) concentrations in road dusts and roadside soils from Perth, Western Australia. The Science of the Total Environment, 317: 121–135.
59. Yunker MB, Macdonald RW, Vingarzan R, Mitchell RH, Goyette D, Sylvestre S (2002). PAHs in the Fraser River basin: a critical appraisal of PAH ratios as indicators of PAH source and composition. Organic Geochemistry, 33: 489–515.
60. Meharg AA, Wright J, Leeks GJL, Wass PD, Owens PN, Walling DE, Osborn D (2003). PCB congener dynamics in a heavily industrialised river catchment. The Science of the Total Environment, 314–316: 439–450.
61. Owens PN, Walling DE, Carton J, Meharg AA, Wright J, Leeks GJL (2001). Downstream changes in the transport and storage of sediment-associated contaminants (P, Cr and PCBs)

in agricultural and industrialised drainage basins. The Science of the Total Environment, 266: 177–186.

62. Walling DE, Owens PN, Carter J, Leeks GJL, Lewis S, Meharg AA, Wright J (2003). Storage of sediment-associated nutrients and contaminants in river channels and floodplain systems. Applied Geochemistry, 18: 195–220.
63. Carter J, Walling DE, Owens PN, Leeks GJL (2006). Spatial and temporal variability in the concentration and speciation of metals in suspended sediment transported by the River Aire, Yorkshire, UK. Hydrological Processes, 20: 3007–3027.
64. Sharpley AN, Smith SJ (1990). Phosphorus transport in agricultural runoff: the role of soil erosion. In: Boardman J, Foster IDL, Dearing JA (Eds), Soil Erosion on Agricultural Land. Wiley, Chichester, pp. 351–366.
65. Harrod TR (1994). Runoff, soil erosion and pesticide pollution in Cornwall. In: Rickson RJ (Ed.), Conserving Soil Resources: European Perspectives. CAB International, Wallingford, pp. 105–115.
66. Rowan JS (1995). The erosional transport of radiocaesium in catchment systems: a case study of the River Exe, Devon. In: Foster IDL, Gurnell AM, Webb BW (Eds), Sediment and Water Quality in River Catchments. Wiley, Chichester, pp. 331–351.
67. Haygarth PM, Hepworth L, Jarvis SC (1998). Forms of phosphorus transfer in hydrological pathways from soil under grazed grassland. European Journal of Soil Science, 49: 65–72.
68. Dils RM, Heathwaite AL (2000). Tracing phosphorus movement in agricultural soils. In: Foster IDL (Ed.), Tracers in Geomorphology. Wiley, Chichester, pp. 259–276.
69. Warren N, Allan IJ, Carter JE, House WA, Parker A (2003). Pesticides and other micro-organic contaminants in freshwater sedimentary environments – a review. Applied Geochemistry, 18: 159–194.
70. Deeks LK, McHugh M, Owens PN (2005). Faecal contamination of water courses from farm waste disposal for three sites in the UK with contrasting soil types. Soil Use and Management, 21: 212–220.
71. Owens PN, Walling DE, He Q (1996). The behaviour of bomb-derived caesium-137 fallout in catchment soils. Journal of Environmental Radioactivity, 32: 169–191.
72. Owens PN, Deeks LK (2004). The role of soil phosphorus in controlling sediment-associated phosphorus transfers in river catchments. In: Golosov V, Belyaev V, Walling DE (Eds), Sediment Transfers Through the Fluvial System. IAHS Publication 288. IAHS Press, Wallingford, pp. 444–450.
73. Hardy IAJ, Carter AD, Leeds-Harrison PB, Foster IDL, Sanders RM (2000). The origin of sediment in field drainage water. In: Foster IDL (Ed.), Tracers in Geomorphology. Wiley, Chichester, pp. 244–257.
74. Foster IDL, Chapman AS, Hodgkinson RM, Jones AR, Lees JA, Turner SE, Scott M (2003). Changing suspended sediment and particulate phosphorus loads and pathways in underdrained lowland agricultural catchments: Herefordshire and Worcestershire, UK. Hydrobiologia, 494: 119–126.
75. Laubel AR, Kronvang B, Larsen SE, Pedersen ML, Svendsen LM (2000). Bank erosion as a source of sediment and phosphorus delivery to small Danish streams. In: Stone M (Ed.), The Role of Erosion and Sediment Transport in Nutrient and Contaminant Transfer. IAHS Publication 263. IAHS Press, Wallingford, pp. 75–82.
76. Walling DE, Owens PN (2003). The role of overbank floodplain sedimentation in catchment contaminant budgets. Hydrobiologia, 494: 83–91.
77. Förstner U, Heise S, Shwartz R, Westrich B, Ahlf W (2004). Historical contaminated sediments and soils at the river basin scale: examples from the Elbe river catchment area. Journal of Soils and Sediments, 4: 247–260.

78. Hudson-Edwards KA, Macklin MG, Taylor MP (1999). 2000 years of sediment-borne heavy metal storage in the Yorkshire Ouse basin, NE England, UK. Hydrological Processes, 13: 1087–1102.
79. Gäbler H-E (1997). Mobility of heavy metals as a function of pH of samples from an overbank sediment profile contaminated by mining activities. Journal of Geochemical Exploration, 58: 185–194.
80. Hudson-Edwards KA, Macklin MG, Curtis CD, Vaughan DJ (1998). Chemical remobilisation of contaminant metals within floodplain sediments in an incising river system: implications for dating and chemostratigraphy. Earth Surface Processes and Landforms, 23: 671–684.
81. Brunet RC, Astin KB (2000). A 12-month sediment and nutrient budget in a floodplain reach of the River Adour, southwest France. Regulated Rivers, 16: 267–277.
82. Miller JR, Lechler PJ, Desilets M (1998). The role of geomorphic processes in the transport and fate of mercury within the Carson River Basin, west-central Nevada. Environmental Geology, 33: 249–262.
83. Horowitz AJ, Meybeck M, Idalfkih Z, Biger E (1999). Variation in trace element geochemistry in the Seine River basin based on floodplain deposits and channel bed sediment. Hydrological Processes, 13: 1329–1340.
84. Kronvang B, Laubel A, Larsen SE, Friberg N (2003). Pesticides and heavy metals in Danish streambed sediment. Hydrobiologia, 494: 93–101.
85. Vörösmarty CJ, Meybeck M, Fekete B, Sharma K, Green P, Syvitski JPM (2003). Anthropogenic sediment retention: major global impacts from registered river impoundments. Global and Planetary Change, 39: 169–190.
86. Renwick WH, Smith SV, Bartley JD, Buddemeier RW (2005). The role of impoundments in the sediment budget of the conterminous United States. Geomorphology, 71: 99–111.
87. House WA, Warwick MS (1999). Interactions of phosphorus with sediments in the River Swale, Yorkshire, UK. Hydrological Processes, 13: 1103–1115.
88. Bowes MJ, House WA, Hodgkinson RA (2003). Phosphorus dynamics along a river continuum. The Science of the Total Environment, 313: 199–212.
89. Jarvie HP, Jurgens MD, Williams RJ, Neal C, Davies JJL, Barrett C, White J (2005). Role of river bed sediments as sources and sinks of phosphorus across two major eutrophic UK river basins: the Hampshire Avon and the Herefordshire Wye. Journal of Hydrology, 304: 51–74.
90. Surridge B, Heathwaite L, Baird A (2005). The exchange of phosphorus between riparian wetland sediments, pore waters and surface water. In: Dynamics and Biochemistry of River Corridors and Wetlands. IAHS Publication 294. IAHS Press, Wallingford.
91. Rae JE, Allen JRL (1993) The significance of organic matter degradation in the interpretation of historical pollution trends in depth profiles of estuarine sediment. Estuaries, 16: 678–682.
92. Rivera-Duarte I, Flegal AR (1997). Pore-water silver concentrations and benthic fluxes from contaminated sediments of San Francisco Bay, California, USA. Marine Chemistry, 56: 15–26.
93. Shine JP, Ika R, Ford TE (1998). Relationship between oxygen consumption and sediment-water fluxes of heavy metals in coastal sediments. Environmental Toxicology and Chemistry, 17: 2325–2337.
94. Schumm SA (1977). The Fluvial System. John Wiley and Sons, New York.
95. Kondolf GM (1994). Geomorphic and environmental effects of instream gravel mining. Landscape and Urban Planning, 28: 225–243.
96. Kondolf GM (1997). Hungry water: effects of dams and gravel mining on river channels. Environmental Management, 21: 533–551.

97. Buffle J, De Vitre RR (1994). Chemical and Biological Regulation of Aquatic Systems. Lewis/CRC Press, Boca Raton, FL.
98. Church MA, McLean DG, Wolcott JF (1987). River bed gravels: sampling and analysis. In: Thorne CR, Bathurst JC, Hey RD (Eds), Sediment Transport in Gravel-Bed Rivers. John Wiley and Sons, New York, pp. 269–325.
99. Church MA (1978). Paleo-hydrological reconstruction from a Holocene valley fill. In: Miall AD (Ed.), Fluvial Sedimentology. Canadian Society of Petroleum Geologists, 5: 743–772.
100. Kondolf GM, Wilcock PR (1996). The flushing flow problem: defining and evaluating objectives. Water Resources Research, 32: 2589–2599.
101. Buffington JM, Montgomery DR (1997). A systematic analysis of eight decades of incipient motion studies, with special reference to gravel-bedded rivers. Water Resources Research, 33: 1993–2029.
102. Shields A (1936). Anwendung der Aehnlichkeitsmechanik und der Turbulenzforschung auf die Geschiebebewegung. Mitteilung der Preussischen Versuchsanstalt für Wasserbau und Schiffbau, 26, Berlin.
103. Baker VR, Ritter DF (1975). Competence of rivers to transport coarse bedload material. Geological Society of America Bulletin, 86: 975–978.
104. Carling PA (1983). Threshold of coarse sediment transport in broad and narrow natural streams. Earth Surface Processes and Landforms, 8: 1–18.
105. Hammond FD, Heathershaw AD, Langhorne DN (1984). A comparison between Shields' threshold criterion and the movement of loosely packed gravel in a tidal channel. Sedimentology, 31: 51–62.
106. Milhous RT (1973). Sediment transport in a gravel-bottomed stream. PhD thesis. Oregon State University, Corvallis, OR.
107. Parker G, Klingeman PC, McLean DG (1982). Bedload and size distribution in gravel-bed streams. ASCE Journal of Hydraulics Division, 108: 544–571.
108. Andrews ED (1983). Entrainment of gravel from naturally sorted riverbed material. Geological Society of America Bulletin, 94: 1225–1231.
109. Misri RL, Garde RJ, Ranga Raju KJ (1984). Bed load transport of coarse non-uniform sediments. ASCE Journal of Hydraulics Division, 110: 312–328.
110. Komar PD (1987). Selective grain entrainment by a current from a bed of mixed sizes: a reanalysis. Journal of Sedimentary Petrology, 57: 203–211.
111. Reid I, Frostick LE (1994). Fluvial sediment transport and deposition. In: Pye K (Ed.), Sediment Transport and Depositional Processes. Blackwell, Oxford, pp. 89–155.
112. Simons DB, Senturk F (1977). Sediment Transport Technology. Water Resources Publications, Fort Collins, CO.
113. McPherson HJ (1971). Dissolved, suspended and bedload movement patterns in Two O'clock Creek, Rocky Mountains, Canada, Summer, 1969. Journal of Hydrology, 12: 221–233.
114. Dietrich W, Dunne T (1978). Sediment budget for a small catchment in mountainous terrain. Zeitschrift für Geomorphologie, N.F. Suppl. Bd, 29: 191–206.
115. Dietrich WE, Gallinati JD (1991). Fluvial geomorphology. In: Slaymaker O (Ed.), Field Experiment and Measurement Programs in Geomorphology. A.A. Balkema, Rotterdam, pp. 169–220.
116. Hoey T (1992). Temporal variations in bedload transport rates and sediment storages in gravel-bed rivers. Progress in Physical Geography, 16: 319–338.
117. Nicholas AP, Ashworth PJ, Kirkby MJ, Macklin MG, Murray T (1995). Sediment slugs: large scale fluctuations in fluvial sediment transport rates and storage volumes. Progress in Physical Geography, 19: 500–519.

118. Emmett WW (1975). The channels and waters of the Upper Salmon River area, Idaho. United States Geological Survey, Professional Paper 870A.
119. Reid I, Frostick LE, Layman JT (1985). The incidence and nature of bedload transport during flood flows in coarse-grained alluvial channels. Earth Surface Processes and Landforms, 10: 33–44.
120. Gomez B, Naff RL, Hubbell DW (1989). Temporal variations in bedload transport rates associated with the migration of bedforms. Earth Surface Processes and Landforms, 14: 135–156.
121. Hubbell DW (1987). Bedload sampling and analysis. In: Thorne CR, Bathurst JC, Hey RD (Eds), Sediment Transport in Gravel-bed Rivers. John Wiley & Sons, New York, pp. 89–120.
122. Parker A, Bergmann H, Heininger P, Leeks GJL, Old GH (2007). Sampling of sediments and suspended matter. In: Barceló D, Petrovic M (Eds), Sustainable Management of Sediment Resources: Sediment Quality and Impact Assessment of Pollutants. Elsevier, Amsterdam, pp. 1–34.
123. Walling DE, Moorehead PW (1989). The particle size characteristics of fluvial suspended sediment: an overview. Hydrobiologia, 176–177: 125–149.
124. Walling DE, Owens PN, Waterfall BD, Leeks GJL, Wass PD (2000). The particle size characteristics of fluvial suspended sediment in the Humber and Tweed catchments, UK. The Science of the Total Environment, 251–252: 205–222.
125. Droppo IG (2001). Rethinking what constitutes suspended sediment. Hydrological Processes, 15: 1551–1564.
126. Petticrew EL (2005). The composite nature of suspended and gravel stored fine sediment in streams: a case study of O'Ne-eil Creek, British Columbia, Canada. In: Droppo IG, Leppard GG, Liss SN, Milligan TG (Eds), Flocculation in Natural and Engineered Environmental Systems. CRC Press, Boca Raton, FL, pp. 71–93.
127. Phillips JM, Walling DE (2005). Intra-storm and seasonal variations in the effective particle size characteristics and effective particle density of fluvial suspended sediment in the Exe basin, Devon, United Kingdom. In: Droppo IG, Leppard GG, Liss SN, Milligan TG (Eds), Flocculation in Natural and Engineered Environmental Systems. CRC Press, Boca Raton, FL, pp. 47–70.
128. Fowler SW, Knauer GA (1986). Role of large particles in the transport of elements and organic compounds through the oceanic water column. Progress in Oceanography, 16: 147–194.
129. Kretzschmar R, Borkovec M, Grolimund D, Elimelech M (1999). Mobile subsurface colloids and their role in contaminant transport. Advances in Agronomy, 66: 121–193.
130. Heathwaite L, Haygarth P, Matthews R, Preedy N, Butler P (2005). Evaluating colloidal phosphorus delivery to surface waters from diffuse agricultural sources. Journal of Environmental Quality, 34: 287–298.
131. Horowitz AJ, Rinella FA, Lamothe P (1990). Variation in suspended sediment and associated trace element concentrations in selected riverine cross sections. Environmental Science and Technology, 24: 1313–1320.
132. Droppo IG, Jaskot C (1995). The impact of river transport characteristics on contaminant sampling error and design. Environmental Science and Technology, 28: 161–170.
133. Gibbs RJ (1977). Transport phases of transition metals in the Amazon and Yukon rivers. Geological Society of America Bulletin, 88: 829–843.
134. Lewin J, Macklin MG (1987). Metal mining and floodplain sedimentation in Britain. In: Gardiner V (Ed.), International Geomorphology 1986, Part 1. Wiley, Chichester, pp. 1009–1027.
135. Westrich B (2004). Personal communication.

136. Old GH, Leeks GJL, Packman JC, Smith BPG, Lewis S, Hewitt EJ, Holmes M, Young A (2003). The impact of a convectional summer rainfall event on river flow and fine sediment transport in a highly urbanised catchment: Bradford, West Yorkshire. The Science of the Total Environment, 314–316: 495–512.
137. Langbein WB, Schumm SA (1958). Yield of sediment in relation to mean annual precipitation. Transactions of the American Geophysical Union, 39: 1076–1084.
138. Walling DE, Kleo AHA (1979). Sediment yields of rivers in areas of low precipitation: a global view. In: The Hydrology of Areas of Low Precipitation, IAHS Publication 128, IAHS Press, Wallingford, pp. 479–493.
139. Lewis SG, Maddy D, Scaife RG (2001). The fluvial system response to abrupt climate change during the last cold stage: the Upper Pleistocene River Thames fluvial succession at Ashton Keynes, UK. Global and Planetary Change, 28: 341–359.
140. Goodbred SL Jr (2003). Response of the Ganges dispersal system to climate change: a source-to-sink view since the last interstade. Sedimentary Geology, 162: 83–104.
141. Semadeni-Davies A (2004). Urban water management vs climate change: impacts on cold region waste water inflows. Climatic Change, 64: 103–126.
142. Macklin MG, Lewin J (1989). Sediment transfer and transformation of an alluvial valley floor: the River South Tyne, Northumbria, UK. Earth Surface Processes and Landforms, 14: 232–246.
143. Miller JR (1997). The role of fluvial geomorphic processes in the dispersal of heavy metals from mine sites. Journal of Geochemical Exploration, 58: 101–118.
144. Walling DE, Probst J-L (Eds) (1997). Human Impact on Erosion and Sedimentation. IAHS Publication 245. IAHS Press, Wallingford.
145. Stone M (Ed.) (2000). The Role of Erosion and Sediment Transport in Nutrient and Contaminant Transfer. IAHS Publication 263. IAHS Press, Wallingford.
146. Golosov V, Belyaev V, Walling DE (Eds) (2004). Sediment Transfer through the Fluvial System. IAHS Publication 288. IAHS Press, Wallingford.
147. Bortone G, Palumbo L (Eds) (2007). Sustainable Management of Sediment Resources: Sediment and Dredged Material Treatment. Elsevier, Amsterdam.
148. Stone KE, Kaufman BS (1985). Sand rights, a legal system to protect the shores of the beach. In: McGrath J (Ed.), California's Battered Coast. Proceedings from a Conference on Coastal Erosion. California Coastal Commission, pp. 280–297.
149. Fan S, Springer FE (1993). Major sedimentation issues at the Federal Energy Regulatory Commission. In: Fan S, Morris G (Eds), Notes on Sediment Management in Reservoirs. Water Resources Publications, Colorado, pp. 1–8.
150. Avendaño C, Cobo R, Sanz ME, Gómez JL (1997). Capacity situation in Spanish reservoirs. I.C.O.L.D. 19th Congress on Large Dams, 74, R.52, pp. 849–862.
151. Batalla RJ (2003). Sediment deficit in rivers caused by dams and instream gravel mining. A review with examples from NE Spain. Cuaternario y Geomorfología, 17: 79–91.
152. Sanz ME, Avendaño C, Cobo R (1999). Influencia de los embalses en el transporte de sedimento hasta el delta del Ebro (España). Proceedings of the Hydrological and Geochemical Processes in Large Scale River Basins Symposium, Manaus, Brasil, pp. 1–6.
153. Vericat D, Batalla, RJ (2006). Sediment transport in a large impounded river: the lower Ebro, NE Iberian Peninsula. Geomorphology, 79: 72–92.
154. Guillén J, Díaz JI, Palanques A (1992). Cuantificación y evolución durante el siglo XX de los aportes de sedimento transportado como carga de fondo por el río Ebro al medio marino. Revista de la Sociedad Geológica de España, 5: 27–37.
155. Palanques A, Drake D (1990). Distribution and dispersal patterns of suspended particulated matter on the Ebro continental shelf, Northwestern Mediterranean. Marine Geology, 95: 193–206.

156. Rovira A, Batalla RJ, Sala M (2005). Response of a river sediment budget after historical gravel mining. The Lower Tordera (NE Spain). River Research and Applications, 21: 829–847.
157. McCulloch M, Fallon S, Wyndham T, Hendy E, Lough J, Barnes D (2003). Coral record of increased sediment flux to the inner Great Barrier Reef since European settlement. Nature, 421: 727–730.
158. Dearing JA, Jones RT (2003). Coupling temporal and spatial dimensions of global sediment flux through lake and marine records. Global and Planetary Change, 39: 147–168.
159. Foster IDL (2006). Lakes and reservoirs in the sediment delivery system: reconstructing sediment yields. In: Owens PN, Collins AJ (Eds), Soil Erosion and Sediment Redistribution in River Catchments: Measurement, Modelling and Management. CABI, Wallingford, pp. 128–142.
160. Foster IDL, Lees JA (1999). Changes in the physical and chemical properties of suspended sediment delivered to the headwaters of the LOIS river basins over the last 100 years: a preliminary analysis of lake and reservoir bottom sediment. Hydrological Processes, 13: 1067–1086.
161. Wolman MG, Shick AP (1967). Effects of construction on fluvial sediment, urban, and suburban areas of Maryland. Water Resources Research, 3: 451–464.
162. Ponczynski JJ, Abrahams AD (1983). Impact of construction activities on sediment response in a small drainage basin, Western New York. Physical Geography, 4: 25–37.
163. Batalla RJ, Sala M (1996). Impact of land use practices on the sediment yield of a partially disturbed Mediterranean catchment. Zeitschrift für Geomorphologie, Suppl. Bd, 107: 79–93.
164. Tiedemann AR, Conrad CE, Dietrich JH, Hornbeck JW, Meghan WF, Viereck LA, Wade DD (1979). Effects of fire on water. A state-of-knowledge review. General Technical Report WO-10. USDA, Forest Service, Washington, DC.
165. Blake WH, Wallbrink PJ, Doerr SH, Shakesby RA, Humphreys GS (2004). Sediment redistribution following wildfire in the Sydney region, Australia: a mineral magnetic tracing approach. In: Golosov V, Belyaev V, Walling DE (Eds), Sediment Transfer Through the Fluvial System. IAHS Publication 288. IAHS Press, Wallingford, pp. 52–59.
166. Blake WH, Wallbrink PJ, Doerr SH, Shakesby RA, Humphreys GS (2006). Magnetic enhancement in wildfire-affected soil and its potential for sediment source-ascription. Earth Surface Processes and Landforms, 31: 249–264.
167. Owens PN, Blake WH, Petticrew EL (2006). Changes in sediment sources following wildfire in mountainous terrain: a paired catchment approach, British Columbia, Canada. Water, Air and Soil Pollution: Focus, 6: 637–645.
168. Van Wyk DB (1982). Influence of prescribed burning on nutrient budgets of mountain fynbos catchments in the south-western Cape, Republic of South Africa. Proceedings of the Symposium on Dynamics and Management of Mediterranean Type Ecosystems. USDA Forest Service General Technical Report PSW-58. Pacific Southwest Forest and Range Experiment Station, Berkeley, California, pp. 390–396.
169. Megahan WF, King JG, Seyedbagheri KA (1995). Hydrologic and erosional responses of a granitic watershed to helicopter logging and broadcast burning. Forest Science, 41: 777–795.
170. Inbar M, Tamir M, Wittenberg L (1998). Runoff and erosion processes after a forest fire in Mount Carmel, a Mediterranean area. Geomorphology, 24: 17–33.
171. Soler M, Sala M, Gallart F (1994). Post fire evolution of runoff and erosion during an eighteen month period. In: Sala M, Rubio JL (Eds), Soil Erosion as a Consequence of Forest Fires. Geoforma Ediciones, Logroño, pp. 149–161.

172. Batalla RJ (2002). Hydrological implications of forest fires: an overview. In: Pardini G, Pinto J (Eds), Fire, Landscape and Biodiversity. Diversitas, Universitat de Girona, pp. 99–116.
173. Brown JAH (1972). Hydrologic effects of a bushfire in a catchment in south-eastern New South Wales. Journal of Hydrology, 15: 77–96.
174. Petticrew EL, Owens PN, Giles T (2006). Wildfire effects on the composition and quantity of suspended and gravel stored sediments. Water, Air and Soil Pollution: Focus, 6: 647–656.
175. Garcia C, Batalla RJ (Eds) (2004). Catchment Dynamics and River Processes: Mediterranean and Other Climate Regions. Elsevier, Amsterdam.
176. Owens PN, Collins AJ (Eds) (2006). Soil Erosion and Sediment Redistribution in River Catchments: Measurement, Modelling and Management. CABI, Wallingford.
177. Perry C, Taylor KG (Eds) (2007). Environmental Sedimentology. Blackwell, Oxford.
178. Van der Perk M (2006). Soil and Water Contamination: From Molecular to Catchment Scale. Taylor & Francis/Balkema, Leiden.

Sustainable Management of Sediment Resources: Sediment Management at the River Basin Scale
Edited by Philip N. Owens

Decision Support Tools for Sediment Management

Marcel van der Perk[a], William H. Blake[b] and Marc Eisma[c]

[a]*Department of Physical Geography, Utrecht University, P.O. Box 80115, 3508 TC Utrecht, The Netherlands*
[b]*School of Geography, University of Plymouth, Plymouth, Devon PL4 8AA, UK*
[c]*Port of Rotterdam, P.O. Box 6622, 3002 AP Rotterdam, The Netherlands*

1. Introduction

Decision support tools assist decision-makers in identifying relevant issues and taking appropriate actions. In the context of sediment management at the river basin scale, these issues relate to sediment quantity or quality. Within the conceptual framework(s) for river basin scale sediment management presented in Chapter 2 (this book), decision support tools aid in the prioritization of sites within the river basin, the site-specific assessment of risks, and the evaluation of management options directed towards the attenuation and prevention of future recurrence of issues related to sediment amounts and quality. Such management options can be undertaken both locally at the problem site and elsewhere in the river basin. There is a wide range of tools available for scientists and managers to facilitate the decision-making process. Until now, there have been no off-the-shelf decision support systems available for integrated management of sediment at the river basin scale that account for all aspects and phases of the sequential process of decision-making. For the time being, decision-makers must therefore largely rely on models and methods that focus on separate aspects and steps. The currently available tools can be classified into four main groups:

- monitoring and mapping tools for assembling information on sediment–contaminant levels and dynamics;
- mathematical models for predicting transport and fate of sediment and contaminants;
- risk assessment tools for the assessment of site-specific risks; and
- cost-benefit analysis and related tools for the evaluation of management options.

The information obtained from each of these groups of tools is different and complementary. This chapter focuses mainly on the first two groups mentioned above: maps and monitoring data; and mathematical models. These tools are particularly directed towards the site prioritization within river basins and the evaluation of options for sediment management. Heise [1] provides further information of risk assessment approaches for sediment management. Chapter 6 (this book) goes into more detail on cost-benefit analysis for the evaluation of sediment management options. Information on sediment treatment tools to clean up contaminated sediment locally can be found in Bortone and Palumbo [2].

The tools discussed in this chapter address the following key questions in sediment management at the river basin scale:

- How much sediment is eroded in the catchment?
- How much of the eroded sediment reaches the river network?
- How much of the sediment transported is deposited in the river network?
- What is the quality of the sediment?
- What is the speciation, mobility and bioavailability of contaminants in sediment?

Given the fact that sediment quantity and quality vary in both space and time, the following additional questions are also important to consider:

- Where do sediments and associated contaminants come from?
- Where do they end up?
- When do erosion and deposition processes occur?
- How does sediment and contaminant transport respond to daily and seasonal climatic conditions?
- Are there long-term trends in sediment transport and sediment quality?

The processes behind these questions are described in Chapter 4 (this book). The following sections aim to provide a broad overview of existing tools that provide answers to one or more of the abovementioned questions. They will successively discuss mapping and monitoring tools, techniques for sediment tracing and fingerprinting, and mathematical models, and will primarily focus on the opportunities and limitations of employing information from measurements and models in answering these questions. However, they do not intend to give a synopsis of general and advanced statistical techniques for data analysis, such as geostatistical interpolation and simulation, multi-criteria analysis, or Pareto analysis. For further background on these methods, we refer to standard texts (e.g. [3–5]).

2. Mapping and monitoring

2.1. Introduction

Environmental information obtained from mapping and monitoring is one of the most widely used tools to support decision-making and to evaluate the effects of management decisions in water and sediment management and control. Mapping is a particularly useful tool for analysing and summarizing information with a dominating spatial component, whereas monitoring is directed towards the temporal component. Water and sediment monitoring records are especially useful for detecting trends in sediment amounts and quality. However, because most water quality monitoring programmes only started in the 1970s and 1980s, long-term historical trends are usually reconstructed using sediment cores from floodplains, lakes and estuaries (e.g. [6–9]). Mapping and monitoring sediment amounts and quality rely on high-quality sampling and measurement. A detailed discussion of different methods of sediment sampling, analysis and assessment is given in Barceló and Petrovic [10].

By the late 1980s and early 1990s, the advent of geographical information systems (GIS) allowed the storage of maps and monitoring data in digital format. Besides the capacity of GIS to store large amounts of spatially distributed environmental data, the effectiveness of GIS as a decision support tool is primarily attributed to the ability to integrate, analyse and visualize these data [11]. As most environmental variables, including sediment transport and sediment quality, vary in both space and time, information from both sources is often combined in spatio-temporal decision support systems that enable the user to analyse and visualize the dynamic properties of the environment. In recent years, along with the rapid development of the Internet, environmental information has been made increasingly accessible to the user (scientists, decision-makers and the general public) through the Internet in the form of on-line databases and Web-GIS. These developments have not only improved the effective dissemination of environmental information, but also have stimulated the role of public participation in the decision-making process [12].

This section provides a brief (and certainly not complete) overview of the sources, analysis and interpretation of environmental information from mapping and monitoring in sediment management. Special attention is paid to the calculation of loads or fluxes of suspended sediment and other substances. Mapping and monitoring data can also be used for input and evaluation of water and sediment quality models. Requirements concerning these data for use in mathematical modelling will be discussed in Section 4 of this Chapter.

2.2. Spatial environmental information

For sediment management at the river basin scale, adequate spatial environmental information on catchment characteristics is essential. This information may include data on physical geographical characteristics (e.g. catchment delineation, elevation, geology and geochemistry, soil type, land cover and land use, and hydrology), diffuse and point-source emissions of contaminants, and monitoring data (e.g. river discharge, water quality, sediment composition and quality, and thickness of sediment layers). Spatial data layers on the physical geographical catchment characteristics and contaminant emissions are generally used as input for models. Emission data are usually either based on direct measurements (mostly only point-source emissions) or census data on population, industry and agriculture. Monitoring data can be employed for site prioritization and site-specific assessment by either ranking of indicator factors or process-based modelling. Furthermore, they can also be used for model calibration and validation.

Spatial environmental information can be derived from various sources. These sources can be grouped according to the method of data acquisition:

- field observation and sampling (e.g. soil type, land use, monitoring data);
- remote sensing (e.g. land cover, elevation);
- census data (e.g. population, agriculture, industry); and
- output from models (e.g. water quality models, soil erosion models, sediment transport models).

Table 1 shows an overview of some generic, freely available European and global geospatial data sets. These data sets primarily represent information on physical geographical catchment characteristics, but also some model output from a regional soil erosion model. These data sets are adequate for analysis and modelling at the scale of large river basins. More spatially detailed information for more local-scale purposes is available for most countries, but these data are often available for restricted use only. Licence conditions and costs vary per country, data set and type of use (e.g. commercial, non-commercial, academic). Furthermore, census data and other environmental statistics can be available from national statistical or environment agencies.

2.3. Water and sediment quality monitoring

Traditionally, environmental information on sediment quantity and quality is derived from regular monitoring programmes or occasional measurement

Table 1. Generic European and global spatial data sets

Data type	Name	Resolution /scale	Producer	WWW
Digital Elevation Model (DEM)	Shuttle Radar Topography Mission (SRTM) (2000)	90 m	NASA	http://www2.jpl.nasa. gov/ srtm/
	GTOPO30 (1996)	30 arc sec (1 km^2)	USGS	http://edc.usgs.gov/products/el evation/gtopo30/ gtopo30.html
Rivers and catchments	European rivers and catchments database (ERICA) (1998)	1:1,000,000	EEA	http://dataservice.eea. europa.eu/dataservice/
	River and catchment database for Europe (CCM) (2003)	1:500,000	EC-JRC	http://agrienv.jrc.it/ activities/catchments/
Land cover	Corine land cover 2000 seamless vector database (CLC2000)	1:100,000	European Topic Centre on Terrestrial Environment	http://dataservice.eea. europa.eu/dataservice/
	Global land cover (GLC2000)	1 km^2	EC-JRC-GVM	http://www-gvm.jrc.it/ glc2000/
Soil type	Corine European soil database version 2	1:1,000,000	EEA	http://dataservice.eea. europa.eu/dataservice/
	European soil database	1:1,000,000/ 1 km^2	EC-JRC-IES	http://eusoils.jrc.it/data.html
Geo-chemistry (soil, sediment)	FOREGS Geochemical Baseline Programme (FGBP)	160 km^2	Geological Survey of Finland (GTK)	http://www.gsf.fi/publ/ foregsatlas/
Soil erosion	Pan-European Soil Erosion Risk Assessment for Europe (PESERA)	1 km^2	EC-JRC	http://eusoils.jrc.it/ ESDB_Archive/pesera/ pesera_data.html

campaigns aimed at water quality assessment, provided that sediment concentrations and sediment quality are measured. Information from water quality monitoring is useful because:

- it allows the assessment of sediment quality;
- it allows the quantification of sediment and contaminant fluxes within and from the river basin;
- it allows the identification of temporal trends in system behaviour and response;
- it provides baseline values;
- it helps to identify the source(s) of the sediment and contaminants; and
- it helps to understand how the sediment–contaminant system behaves and functions at a variety of scales from particle interactions to the basin scale.

Table 2 gives an overview of the various types of water quality assessment and their objectives.

Table 2. Objectives of water quality assessment operations (source: after [14])

Type of monitoring		Major focus
Regular monitoring		
1	Multipurpose monitoring	Spatial and temporal distribution of water quality
2	Trend monitoring	Long-term trends in concentrations and loads
3	Basic survey	Identification of location and spatial distribution of major survey problems
4	Operational surveillance	Water quality for specific uses
Occasional monitoring		
5	Baseline monitoring	Baseline levels and their spatial variability; used as reference point for pollution and impact assessments
6	Preliminary survey	Inventory of pollutants and their spatio-temporal variability prior to monitoring programme design
7	Emergency surveys	Rapid assessment of pollution following a catastrophic event
8	Impact surveys	Sampling near pollution sources limited in time and space, generally focusing on few variables
9	Modelling surveys	Intensive water quality assessment limited in time and space and choice of variables for calibration/validation of water quality models
10	Early warning surveillance	Continuous and sensitive measurements at critical water use locations (e.g. drinking water intakes)

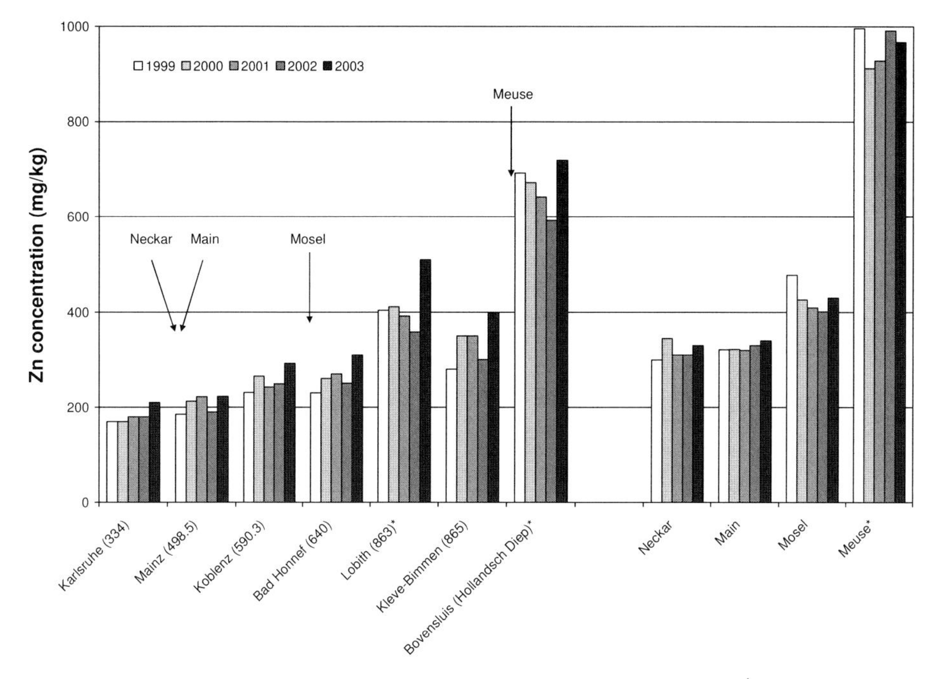

Figure 1. Annual median zinc concentrations in suspended sediment (mg kg^{-1} dry matter) in the River Rhine (including Hollandsch Diep) and major tributaries between 1999 and 2003. The figures in parentheses refer to river kilometres (source: DK Rhein [16] and Rijkswaterstaat [17])

Figure 1 shows an example of results from water quality monitoring programmes in the River Rhine basin. In the German reach from Karlsruhe to Kleve-Bimmen/Lobith, the median zinc concentrations in suspended sediment particularly increase in the last river stretch, where the highly populated Ruhr area is situated. The relatively high median concentrations measured in the Hollandsch Diep near the outflow to the North Sea can be largely attributed to the relatively large metal input from the Meuse River. From Figure 1 there is no consistent trend in the zinc concentrations detectable, but the zinc concentrations are closely negatively related to the mean annual discharge ($r = 0.977$ for the 5 years at Lobith). This explains the relatively high concentrations in 2003, when the mean annual discharge at Lobith was only 1821 $m^3\ s^{-1}$, whereas in the other years the mean annual discharge ranged between 2521 $m^3\ s^{-1}$ (2000) and 2974 $m^3\ s^{-1}$ (2002).

Data on bed sediment quality are much rarer than data on suspended sediment quality. Heise *et al.* [13] provides an extensive assessment of existing data on bed sediments in the River Rhine and its tributaries. They used these data to identify areas which show increased concentrations of contaminants and potentially pose a risk to downstream areas, including the Port of Rotterdam.

It should be noted that monitoring programmes may differ for different countries and monitoring stations. These differences may include differences in sampling frequency and coverage, sampling depth, sample storage and treatment protocols, analytical techniques, and relate to different monitoring objectives, strategies and traditions. Also, data availability may differ. This may give rise to inconsistencies in monitoring data sets, particularly in the case of transboundary river basins. A striking example of this can be seen in Figure 1. The monitoring stations at Kleve-Bimmen (Germany; left bank) and Lobith (the Netherlands; right bank) are located almost opposite to each other. Nevertheless, the median concentrations measured at the Dutch side of the border are on average 24% higher than those measured at the German side of the border. Other issues of uncertainty in monitoring data are discussed by Rode and Suhr [15].

2.4. Calculation of fluxes

The average load of sediment or another substance that has passed a point in a river over a period of time can be defined as:

$$L = \frac{\int_{0}^{t_1} C(t) \cdot Q(t)\, dt}{t_1 - t_0} \qquad (1)$$

where L = the load that passes a point in the time interval from $t = 0$ to $t = t_1$ [M T^{-1}], $C(t)$ = the volume concentration of a substance as function of time t [M L^{-3}], $Q(t)$ = the water discharge as function of time [L^3 T^{-1}]. Because concentration and discharge are often correlated, the annual average load is not equal to the annual average concentration times the annual average discharge. If there is a positive correlation between discharge and concentration due to increased substance inputs during high flow conditions, for instance as a result of soil erosion, the simple product of annual average concentration and discharge underestimates the annual load. Conversely, if there is a negative correlation due to dilution during high flow conditions, the simple product of annual average concentration and discharge overestimates the annual load [18].

To estimate long-term loads of substances in rivers, various methods are used; for an overview of these methods, we refer to [19] and [20]. A long-term discharge–concentration relation is the method adopted most frequently for the prediction of unmeasured substance concentrations from water discharge [21–23]. Such a long-term relationship is called a concentration rating curve. Rating curves have especially been used to estimate sediment concentrations or loads (e.g. [23, 24]). Usually a concentration rating curve takes the form of a power function:

$$C = a\,Q^{b} \qquad (2)$$

where C = the concentration of the substance under consideration (mg l^{-1}), Q = water discharge ($m^3\ s^{-1}$) and a and b are coefficients. Log transformation of Equation 2 yields a linear equation:

$$\log C = \log a + b \log Q \qquad (3)$$

where the logarithms are to base 10. Fitting a line to the log–log plot of C against Q using least squares linear regression provides the values for a and b. Using Equation 2 to predict C, however, yields the geometric – not arithmetic – mean of the statistical distribution of C given a value for Q. The geometric mean is always less than or equal to the arithmetic mean. To estimate the arithmetic mean of C, the following bias correction is used [22]:

$$C = a\,Q^{b}\,e^{2.65\,s^2} \qquad (4)$$

where s^2 = mean square error of the regression Equation 3.

Asselman [24] argued that for sediment rating curves, steep curves with a small value for a and a large value for b are characteristic for river sections with little sediment transport during low discharge. An increase in discharge results in a large increment of suspended sediment concentration. This suggests that during high discharge periods the river's power to erode material is large or that important sediment sources become available. Flat rating curves usually occur in river sections with intensively weathered materials or loose sedimentary deposits that are transportable during low discharges.

Horowitz [23] evaluated the use of sediment rating curves for the estimation of suspended sediment concentrations for subsequent load calculations. He showed that for periods of 20 or more years, accuracies of less than 1% can be accomplished using a single sediment rating curve based on data for the entire period. Furthermore, he demonstrated that hydrologically based sampling allows more accurate estimates of average annual loads to be derived compared to calendar-based sampling. Relatively accurate (errors <±20%) annual suspended sediment fluxes can be obtained from annual sediment rating curves calculated using hydrologically based monthly measurements.

Not only the concentration, but also the quality of suspended matter may vary with discharge as the source areas of sediment vary in time during hydrological events. For example, Figure 2 shows the zinc concentrations in suspended sediment as a function of the discharge of the River Rhine near Lobith in the Netherlands. The zinc concentrations in sediment are highly variable when

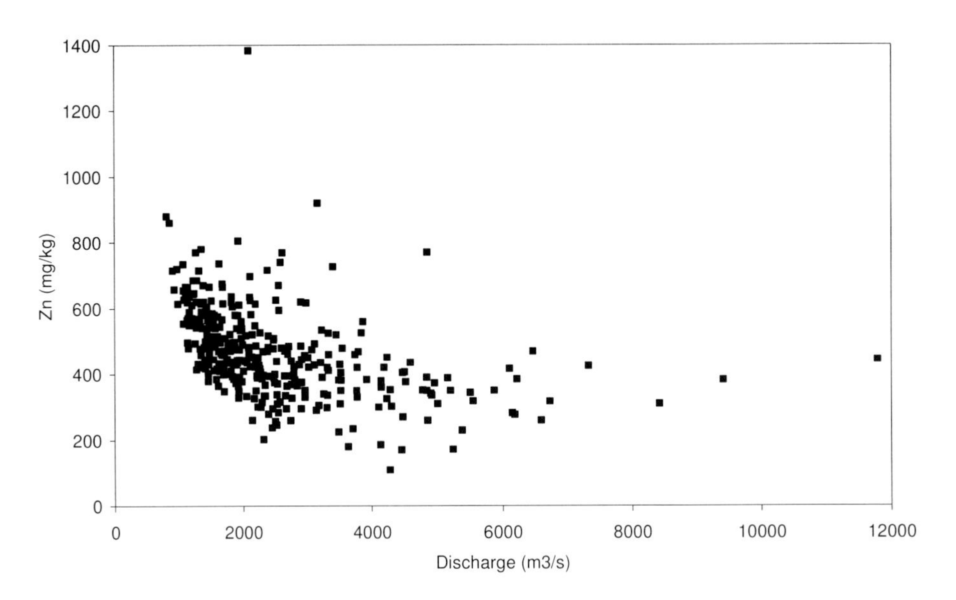

Figure 2. Concentration rating curve for zinc in the River Rhine near Lobith, the Netherlands (1988–2002) (source: [17])

discharge is below about 4000 m^3 s^{-1}. Although few measurements are available for high discharges, the results indicate that when discharge exceeds 4000 m^3 s^{-1}, the concentrations decrease to about one-half to two-thirds of the concentrations during low flow, and show less variation. It seems that during low flow the suspended sediment is highly enriched in contaminants from industrial and domestic point discharges and diffuse urban runoff. During hydrological events, soil erosion on the terrestrial part of the catchment and river bank erosion generate a supply of relatively uncontaminated sediment to the river. Such relations need to be taken into account when calculating average annual loads of sediment-associated contaminants or calculating sediment-associated contaminant deposition on floodplains (see also [25]).

2.5. Source apportionment

When combined with statistical information on industrial and domestic point-source discharges, monitoring data can be used for calculating the contributions from diffuse and point sources to the total contaminant load in the river. The most straightforward approach is to compare the calculated contaminant loads to the total of upstream point-source discharges (immission analysis; see for example [26]). This approach neglects the in-stream contaminant retention and losses due to, for example, interaction with bed sediments, degradation or volatilization (see Chapter 4, this book).

To incorporate the effects of in-stream retention, more sophisticated statistical tools are available. The MESAW-model is such a statistical model for source apportionment of riverine contaminant transport [27]. This model approach uses non-linear regression for the simultaneous estimation of export coefficients (i.e. the mass per unit time per unit area transferred to surface water) for different land use or soil categories, and retention coefficients for pollutants in the river basin. The procedure consists of the following major steps:

1. Estimation of loads at each water quality monitoring site.
2. Subdivision of the entire drainage basin into sub-basins, defined by the monitoring sites for water quality and their upstream–downstream relationships.
3. Collation of statistics on land use, soil type, lake area, point source emissions and other relevant data for each sub-basin.
4. Non-linear regression with loads from each sub-basin as the response variable and load from the upstream sub-basin and sub-basin statistics as explanatory variables to estimate export and retention coefficients.

Related methods of sediment tracing and fingerprinting are discussed in the next section. Sediment tracing and fingerprinting methods often rely on targeted soil and sediment sampling and can, therefore, provide unique and specific information on sources and pathways of sediment and associated contaminants in catchments (see Chapter 4, this book). More advanced mathematical models for the prediction of concentrations and fluxes, which can also be employed for source apportionment, are discussed in Section 4.

3. Sediment tracing and fingerprinting

3.1. Introduction

Field- and laboratory-based sediment tracing technologies offer the opportunity to assess sediment redistribution dynamics in river basins in terms of (i) quantity and (ii) source, thereby answering the key sediment management questions:

- How much material is being mobilized/transported/stored within the system?
- Where is 'problem' sediment coming from?

Tracer techniques have developed considerably over the past three decades since early pioneering work on the use of fallout radionuclides to estimate soil erosion (e.g. [28–30]) and geochemical and mineralogical properties to determine sediment provenance (e.g. [31–33]). Tracing approaches are generally considered to overcome spatial and temporal representation problems associated with more traditional approaches such as erosion pins and plots, and provide complementary information to river sediment monitoring programmes [34]. This section will explore the theoretical framework behind both budget- and source-related sediment tracing approaches and their application in disturbed river basins. Readers are directed to recent review articles for further information [35, 36].

3.2. Estimating sediment redistribution rates using fallout radionuclides

The sediment budget concept [37] is central to tracer approaches that aim to quantify soil erosion, deposition and export from slope environments. The fallout radionuclide approach utilizes readily measurable and ubiquitous environmental radionuclides that are delivered to the soil surface from the atmosphere by both dry and wet precipitation processes. The fallout radionuclides caesium-137 (^{137}Cs), lead-210 (^{210}Pb) and beryllium-7 (^{7}Be) have all been successfully used as tracers in erosion and budgeting studies. Each offers soil and sediment redistribution data at different timescales on account of their different production and delivery dynamics and half-lives. All three isotopes, plus relevant parent or daughter isotopes, are measurable at low levels by high-resolution, low-level gamma spectrometry techniques using high-purity germanium (HPGe) detectors as discussed in detail by Wallbrink *et al.* [38].

To date, anthropogenic ^{137}Cs has been the most widely adopted sediment tracer of these three radioisotopes. Both ^{210}Pb and ^{7}Be approaches are based on the principles of the ^{137}Cs technique. Large quantities of ^{137}Cs were produced in the upper atmosphere during thermonuclear weapons testing in the 1950s and 1960s. The radioisotope was globally distributed in the upper stratosphere prior to surface deposition, although spatial variability in total activity per unit area (i.e. local inventories) have been documented [39]. Caesium-137 has a half-life of 30.2 years, so the activity delivered to soil during the weapons-testing era is readily measurable in surface soil today. In contrast, ^{210}Pb is a naturally occurring radioisotope, part of the uranium-238 (^{238}U) decay series, with a half-life of 22.3 years. Its parent radioisotope, gaseous radon-222 (^{222}Rn), is sourced from the decay of radium-226 (^{226}Ra) in local geological material. The gas diffuses into the atmosphere where it decays to ^{210}Pb, which is subsequently delivered to the surface. Fallout ^{210}Pb is said to be 'unsupported' or termed 'excess ^{210}Pb', since it is not in equilibrium with ^{226}Ra [40]. Beryllium-7 is also naturally occurring, but formed by cosmic ray spallation of O and N in the upper atmosphere. With a half-life of 53.3 days, ^{7}Be is the shortest-lived of the three isotopes, but its delivery dynamics to the surface are similar, i.e. by both wet and dry deposition, although wet deposition dominates [41].

The ^{137}Cs method for measurement of soil and sediment redistribution has several important requirements. First, radionuclide delivery is assumed to be spatially uniform across the study area and, second, once delivered to the surface, radionuclides must be rapidly and irreversibly bound to soil particles [42, 43]. If these assumptions hold true, any subsequent redistribution of radionuclide activity can be related to soil and sediment redistribution in the study area. The assessment of soil redistribution is usually based on comparison of measured inventories (total radionuclide activity per unit area) at specific sampling points of interest to a local 'reference' site that has experienced no erosion or deposition and hence holds an inventory equivalent to the total activity delivered to the study area. This crucial data set is termed the 'reference inventory'. Where sample inventories are lower than the reference, erosion will have occurred, i.e. a proportion of the local ^{137}Cs will have been exported with mobilized soil. Where sample inventories are larger than the reference inventory, deposition will have occurred, i.e. the local area will have received ^{137}Cs from soil delivered from up slope in addition to local fallout inputs. The measured spatial distribution of radionuclides across the study area and local reference inventory can be used to construct a radionuclide budget that, given the above behavioural assumptions, can be converted to a soil and/or sediment budget across the landscape unit of interest. To convert the radionuclide budget to quantitative estimates of soil and sediment redistribution, the relationship between the deviation from the reference inventory and soil erosion or deposition must be clearly defined [29, 44]. Over the past years a variety of calibration models have been developed that convert radionuclide inventories to soil erosion and/or deposition rates. Many models are specific to certain soil conditions (e.g. specific agricultural activity) and furthermore incorporate algorithms to describe potential migration of the radioisotopes in the soil profile. A full appraisal of conversion models is given by Walling *et al.* [45].

Since most ^{137}Cs present in soil, beyond contamination by consented and accidental discharge from the nuclear power industry (notably the Chernobyl accident in 1986), was delivered in the 1950s and 1960s, the contemporary spatial distribution of ^{137}Cs generally represents soil redistribution over the past ca. 50 years. This varies across the globe in line with variation in peak fallout patterns during weapons testing. Hence, redistribution rates derived from the ^{137}Cs approach can be considered a medium-term average rate. Although ^{210}Pb has a shorter half-life, its natural source means that it has been continually produced and delivered to the soil surface. Since five half-lives is the limit for the detection of a radioisotope once produced, spatial patterns in unsupported ^{210}Pb provide soil redistribution rates over the past ca. 100 years, i.e. the longer-term. Beryllium-7 offers the shortest-term perspective with its half-life of 53.3 days. Although some workers have demonstrated its value in providing soil redistribution rates on a single event scale (e.g. [46, 47]), the short-term delivery and redistribution dynamics associated with this radioisotope means that antecedent delivery dynamics must be quantified if the event in question is not isolated within a five half-life time period.

The fallout radioisotope approach may be applied at different spatial scales depending upon the questions being asked by land and river managers and the constraints imposed by the key assumptions discussed above. Ritchie and Ritchie [48] provide a comprehensive list of ^{137}Cs and related studies undertaken up to 2007. Here, two examples are described to illustrate the different degrees of temporal and spatial resolution that can be achieved depending upon key research questions and the limitations of the study environment.

Walling *et al.* [49] undertook high resolution spatial sampling of soil at the intersections of a 10 m grid within a 6.7 ha cultivated field in southwest England. Samples were analysed for ^{7}Be, ^{137}Cs and ^{210}Pb and conversion models applied to produce estimates of soil redistribution. Figure 3a illustrates the spatial resolution in soil erosion data that can be achieved for a relatively small plot allowing erosion and sediment retention information to be linked to small-scale changes in agricultural impacts and field topography. The work showed that the single extreme rainstorm event characterized by ^{7}Be represented an above average erosion event for the study site, but importantly the sediment delivery ratios for the short and medium term were the same, i.e. the same proportion of eroded sediment is delivered from the slope to the stream network.

Wallbrink *et al.* [50] undertook to quantify soil redistribution on hillslopes following forest harvesting using a ^{137}Cs tracer approach. In this case the authors were interested in the transfer of material between landscape units of contrasting impact by logging activities as opposed to detailed spatial patterns of soil redistribution. Their approach involved characterizing the average inventory of ^{137}Cs and ^{210}Pb for each landscape unit of interest using spatially integrated samples, which has the added value of reduced demand on analytical services. The resulting sediment budget is shown in Figure 3b with clear messages about key sediment source zones and sediment retention in the system. On the basis of this information, catchment managers can more effectively develop remediation strategies and plans.

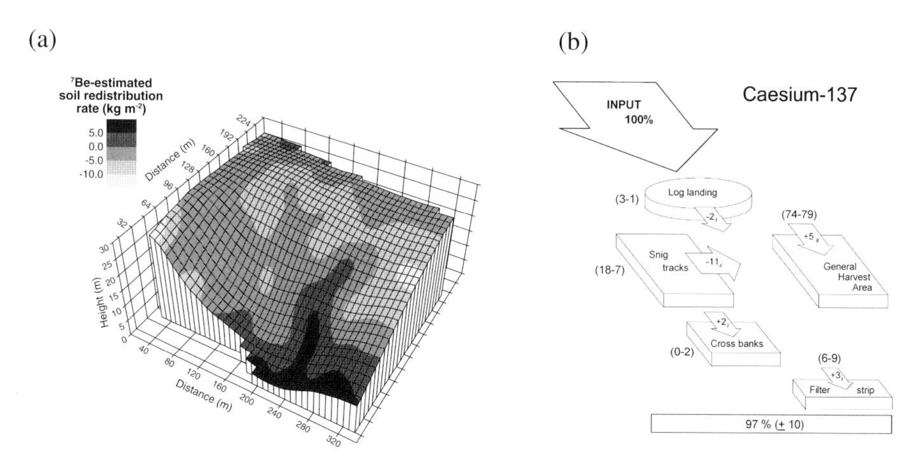

Figure 3. (a) Example of high spatial resolution information on soil redistribution using fallout radionuclides for a field in England (source: [49]); (b) Tracer budget showing areas of erosion and deposition in various landscape elements, for a study in Australia, based on measurements of fallout ^{137}Cs (source: [50] reproduced with permission from Elsevier)

The fallout radionuclide approach can be extended to explore fine sediment residence times in river channels (e.g. [51]) and the sediment column of river basin sediment stores, for example floodplains or lakes, whereby vertical changes in radionuclide concentration can be used to date sediment horizons and infer rates of sediment accumulation (e.g. [52]). Appleby [53] provides a detailed review of current approaches and considerations in dating recent sediment deposits.

3.3. Tracing sediment sources in river basins

Tracer approaches that aim to quantify the role of specific sediment source areas are often termed 'sediment fingerprinting' approaches. These can provide important information on both the spatial source of fine and coarse sediment within the river basin (i.e. specific zones or subcatchments) and the vertical depth in the soil profile from where material has been derived (i.e. surface vs subsurface material or channel bank erosion) (see Figure 4). The latter categorization of source type according to depth provides further information on erosion processes operating within a system, e.g. discriminating between surface wash and the formation of rills or gullies whereby material is mobilized from different depths in the soil profile ([54], also see Chapter 4, this book).

The source-tracing technique relies on the ability to define a unique and traceable signature for potential source materials in a river basin using physical or chemical sediment properties. If suitable discriminating properties are found

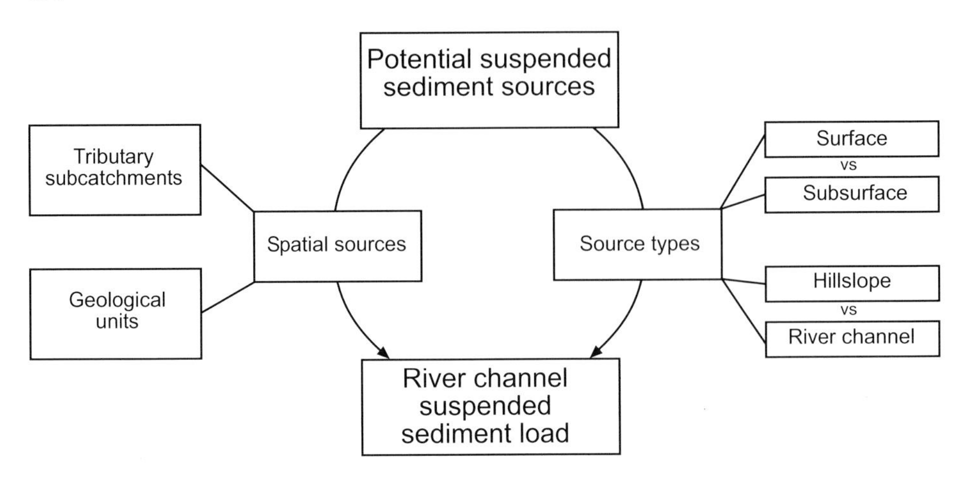

Figure 4. An elementary classification of potential sediment sources in river basins (source: adapted from [55])

in source material, measurement of the same properties in a suspended sediment sample or sedimentary sequence can be undertaken and the resulting signatures compared to the suite of potential source materials. An important assumption of the technique is that the physico-chemical properties of the sediment are not altered during mobilization, transport or subsequent sedimentary storage. If they are, then it is crucial to account for any transformations in sediment signatures before the comparison between source and downstream sediment is made. A good example is the influence of fluvial sorting on sediment grain size composition which, owing to the relationship between many geochemical properties and particle size (cf. [56]), can have a profound effect on elemental concentrations in soil and sediment material. A range of approaches to correct for particle size differences are used, and are described in more detail by Walling [36].

The collection of samples for source characterization requires care to ensure signatures are both representative and unbiased. Whilst random sampling minimizes the potential for bias, the trade-off is additional analytical work characterizing samples of material that do not present a logical source within the study system. The GIS-based analysis of erosion risk to define areas of high erosion potential to guide source sampling [57] offers an efficient and unbiased approach to source sampling. However, many workers favour field-based observation evidence to guide sample collection. Collection of suspended sediment and sedimentary archives also requires some attention. Details on sediment sampling approaches are discussed in detail in [58].

A variety of sediment properties have been adopted by workers to trace mineral material in river basins. Chapter 4 (this book) discusses the use of various contaminants to elucidate sources and transfer pathways of material. In this section we explore primarily the use of natural sediment properties to elucidate sediment sources and pathways in the context of catchment soil erosion and sediment delivery problems. First, major and minor trace element geochemistry of source material (e.g. [31]) is predominantly associated with underlying geology and soil formation processes, although signatures may be modified by anthropogenic pollution [59]. Geochemical signatures may be used to discriminate spatial sources within a river basin if geology and soil are sufficiently variable. Due to pedogenic processes and industrial contamination, surface and subsurface soil material may show significant differences, hence vertical sources may also be determined, often in conjunction with exploration of spatial sources (cf. [60]). Suites of major and minor trace elements are generally determined by X-ray fluorescence (XRF), which gives a total concentration of both mineral-based and adsorbed elements, or Inductively Coupled Mass Spectrometry (ICP-MS) or Atomic Absorption Spectrometry (AAS) following acid leaching. Sediment *mineralogy*-based signatures (e.g. [61]) are again largely related to geology and weathering regime. Second, mineral magnetic properties (e.g. [62, 63]) of soil and sediment relate to the iron mineralogy of the source material which is sensitive to environmental change. Iron minerals that control the signatures are sourced from both geologic (primary minerals) and secondary processes such as pedogenesis, bacterial activity and burning (see [64] for further details). As such, mineral magnetic properties offer good potential for tracing sediment to both spatial sources in larger basins and generic source types related to depth. Third, in addition to their role in sediment budget studies, fallout radionuclides have been successfully used to discriminate generic sediment sources and processes on account of variability in the shape and depth penetration of their depth profiles in soil (e.g. [54]).

Granger *et al.* [65] note a potential drawback in the use of 'naturally-occurring' tracer properties in that it can be difficult to determine the exact amounts of tracer input to any source area. They suggest that in some instances the artificial introduction of exotic particles can be useful, e.g. fluorescent [66–68] and magnetic beads [69]. More recently, work has explored the potential for using rare earth oxides [70, 71], DNA fragments [72] and dioxin congeners [73]. Although most sediment tracing work to date in fluvial systems has focused on mineral material, stable C and N isotope techniques have been applied more recently to trace organic material back to specific terrestrial, biological and aquatic sources (e.g. [74]). In this context, other useful tracers of organic material include artificial fluorescence and DNA labels [65]. Indeed,

Granger *et al.* [65] note that where application of tracer techniques is required to identify spatial sources of particle-associated nutrients, it is important to target the organic phase of problem sediment since the source of 'carrier' mineral sediment may not represent fully the source of attached nutrients, which can be adsorbed/desorbed en route. Whilst selection of an individual property suite may suit certain catchments (e.g. mineral magnetics in a burnt system), a multivariate or composite fingerprint approach is generally favoured (cf. [75]), allowing unequivocal discrimination between several potential sources [36].

Comparison of source signatures to downstream material may be made qualitatively through use of bivariate and multivariate plots. However, most workers prefer to adopt a quantitative (un)mixing model approach, whereby the proportional contribution of each sediment source is estimated by an iterative mixing model algorithm (e.g. [62, 63]). Recent work [76] has highlighted the importance of estimating the uncertainty on derived source proportions, especially where it has been difficult to constrain source signatures. Other foci of study are evaluation of the conservative behaviour of tracer properties during transport and storage, the enrichment of signatures due to changing particle size assemblages and organic matter content (e.g. [77]).

Sediment fingerprinting procedures have been applied at a range of catchment scales in both agricultural and urban catchments. In addition to the sampling of discrete sediment samples, the analysis of lake and floodplain sediment cores can be used to derive a longer term temporal perspective on changing river basin sediment sources [78].

A good example of sediment source-tracing and its role in directing management of catchment resources is illustrated by Evans *et al.* [79] for the River Bush, in Northern Ireland, UK. Here the EU Water Framework Directive has forced regulatory authorities to examine directly the links between catchment erosion processes and downstream water and habitat quality since the river is an important salmon habitat. Pressures within the system include: (i) forestry operations, where clear-felled zones are currently buffered from the stream network by 5 m riparian zones; (ii) grazing of livestock, where high stocking densities and a lack of stream fencing are notable issues; and (iii) extensive land drainage to improve agricultural land. The study involved identification of key sediment sources within the system through mapping of erosion risk using a GIS approach and field reconnaissance. The mineralogical and mineral magnetic properties of potential sediment source materials were then used to develop fingerprints for sediment tracing purposes. The outcome from the analysis was an appraisal of the relative importance of sediment sources within the system as illustrated by Figure 5, with notable differences in the sources of suspended sediment and bed load.

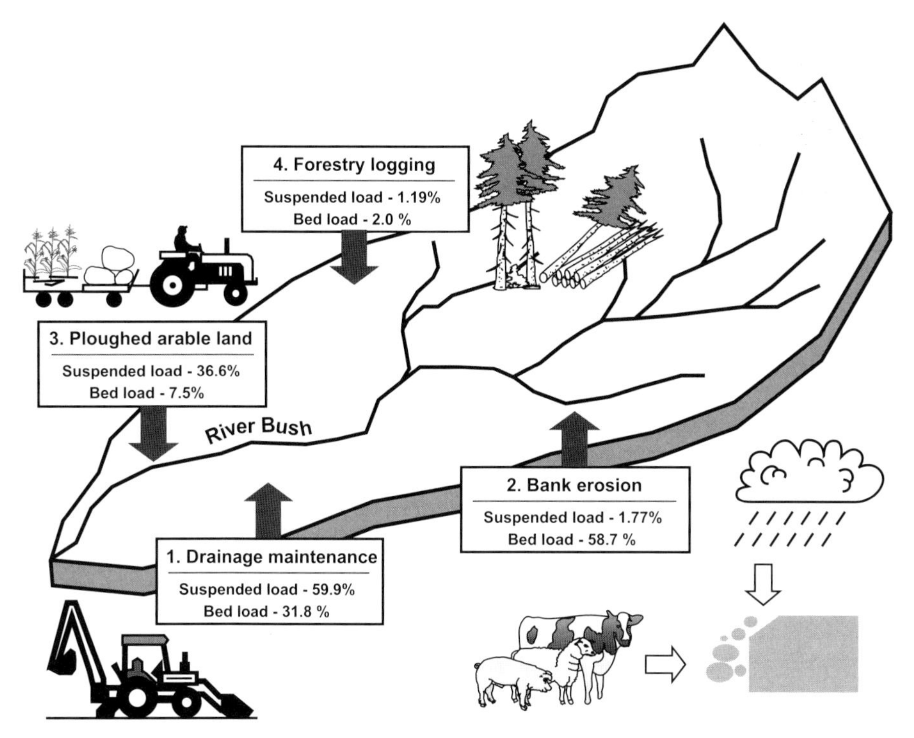

Figure 5. Schematic of load-apportioned sediment sources in the Bush catchment, Northern Ireland (ranked by importance) (source: adapted from [79])

The management implications of the research illustrated in Figure 5 are clear and relate to both minimization of soil erosion and perhaps more importantly reduction of the connectivity between disturbed landscapes and the river network. Recommendations could be made regarding maintenance of drainage channels, use of livestock fencing and further development of riparian buffer zones. However, the management of more diffuse sources of suspended sediment from arable land remains an important challenge which may only be addressed in partnership with the many farmers working within the river basin.

In the context of sediment-related management problems, much attention has focused on the tracing of fine material in river basins as discussed above. A variety of approaches have, however, also been explored for tracing coarser (greater than 4 mm) material within fluvial environments. The coarse load of a river system exerts an important control on habitat quality, changing channel form (erosion and land loss) and deposition impacts on infrastructure [80]. Sear *et al.* [80] classify coarse sediment tracer approaches according to the purpose of their application, namely: (i) determination of particle entrainment thresholds; (ii) description of particle displacement; (iii) particle controls on distance of transport; (iv) morphological controls on transport distance; (v) sediment transport estimation; and (vi) intra-event activity. Although some natural tracer properties have been used, e.g. natural magnetic properties (e.g. [81]), most early applications involved either the alteration of natural material, for example painting individual particles, which relied on visual observation of particle movement, or the introduction of exotic material. More recently, work has focused on magnetic (e.g. [82]) or radio-transmitter inserts (e.g. [83]) to track particles, with greatly improved recovery rates. Further information is available in the detailed review by Sear *et al.* [80].

4. Mathematical models

4.1. Introduction

Mathematical models are powerful tools in sediment management, primarily because they have the potential to predict future conditions under various scenarios of environmental change and management strategies. Furthermore, the employment of models may give the opportunity to achieve a reduction of monitoring costs by replacing or supplementing expensive measurements by less expensive model predictions. The use of computer models for decision support is made or broken by their user-friendliness in terms of model construction, and pre- and post-processing of the data including visualization of the model output. That is why most contemporary computer models have user-friendly graphical user interfaces, whether or not integrated in a GIS. Nevertheless, modelling often remains a complex activity and requires experience and knowledge of the underlying principles and associated uncertainties for an adequate model construction and interpretation of the model outcomes.

There are numerous models available for predicting the sediment dynamics and quality in river basin systems, including statistical models, conceptual models and physically based models. They can be applicable to small catchments of a few hectares to large river basins, and may have temporal resolutions from a few seconds to years. There are one-dimensional models (mostly for the fluvial part of the system), two-dimensional models (mostly for land surface processes such as soil erosion, or for prediction of sediment–contaminant dynamics in reservoirs, floodplain sections and harbours), and three-dimensional models (mostly for local interactions between sediment transport and morphological development or sediment–contaminant dynamics in stratified water bodies, such as deep lakes or estuaries).

This section provides an overview of the various available tools and submodels for simulation of sediment and contaminant behaviour in the river basin continuum from hillslopes, via the surface water network to estuaries and coastal seas. It focuses on the various model approaches, including the spatial and temporal scales they operate at, and on how these model tools can provide answers to the key questions stated in Section 1. For a discussion of the processes (sediment and contaminant transport and pathways) that form the basis for these models, see Chapter 4 of this book.

4.2. Erosion and sediment delivery models

Most erosion models have been developed for field or small catchment scale applications, initially as aids to soil conservation planning [84]. In general, these models describe sheet (interrill) and rill erosion processes, as they are the dominant erosion processes on agricultural hillslopes. Some models also account for gully erosion, but other mechanisms of sediment production, for example landslides, debris flows and river bank erosion, which may be important sediment sources in mountainous catchments, are generally neglected.

Soil erosion models can be grouped according to the temporal scale at which they operate: event-based erosion models estimate soil erosion dynamics during single or successive rainstorm events, whereas long-term erosion models estimate mean annual soil losses. Table 3 lists some common soil erosion models operating at different spatial and temporal scales. These and other models are described in further detail by [85] and [86]. Models for predicting event-scale soil erosion are fully dynamic and mostly based on physical process descriptions of runoff generation and sediment detachment and transport, whereas long-term models are often steady state and based on generalized concepts. It is clear that, accordingly, event-scale models have more demanding data requirements than models concerned with long-term soil erosion rates or risks. Nevertheless, all models require climate data and spatially distributed data on topography (digital elevation model derivatives such as slope, slope length, upstream catchment area; see [11]), soil and vegetation. The USLE and RUSLE models represent the least complex modelling approach, but they do not account for transport capacity-limited sediment transport. Thus, these models do not consider sediment deposition due to local exceedance of the sediment transport capacity, which is a function of overland flow depth and velocity, slope steepness and the transportability of the soil particles. Therefore, they are only capable of estimating soil loss at field-size hillslope sections. Other long-term erosion models, such as the Morgan, Morgan and Finney model (MMF) and SEDEM, do account for the sediment transport capacity of overland flow.

Table 3. Examples of soil erosion models

Acronym	Name	Temporal scale	Temporal resolution	Spatial extent	Source
EUROSEM	European Soil Erosion Model	Event	Minute	Fields/small catchments	[87]
	http://www.silsoe.cranfield.ac.uk/nsri/research/erosion/eurosem.htm				
LISEM	LImburg Soil Erosion Model	Event	Minute	Fields/small catchments	[88]
	http:// www.geog.uu.nl/lisem/				
ANSWERS -2000	Areal Non-point Source Watershed Environment Response Simulation	Event–months	Minute	Fields/small catchments	[89]
	http://www.bse.vt.edu/ANSWERS/				
WEPP	Water Erosion Prediction Project	Event–months	Day	Fields/small catchments	[90]
	http://topsoil.nserl.purdue.edu/nserlweb/weppmain/				
USLE	Universal Soil Loss Equation	Long term	(Steady state)	Plots/hillslope	[91]
RUSLE	Revised Universal Soil Loss Equation	Long term	Day	Fields	[92]
	http://www.ars.usda.gov/Research/				
GLEAMS	Groundwater Loading Effects of Agricultural Management Systems	Long term	Day	Fields	[93]
	http://www.tifton.uga.edu/sewrl/Gleams/gleams_y2k_update.htm				
MMF	Revised Morgan-Morgan-Finney model	Long term	(Steady state)	Fields/small catchments	[94]
SEDEM	Soil erosion and sediment delivery model	Long term	(Steady state)	Regions	[95]
	http://www.kuleuven.be/geography/frg/modelling/erosion/watemsedemhome/				

Table 3 – *continued*.

Acronym	Name	Temporal scale	Temporal resolution	Spatial extent	Source
PSYCHIC	Phosphorus and Sediment Yield CHaracterization In Catchments	Long term	(Steady state)	Catchments/ farm	
	http://www.psychic-project.org.uk/				
PESERA	Pan-European Soil Erosion Risk Assessment	Long term	Month (steady state)	Europe	[96]
	http://eusoils.jrc.it/ESDB_Archive/pesera/pesera_cd/index.htm				

Since most of the erosion models listed in Table 3 tend to be focused on field-scale processes, they may not be well suited for application at broader scales, for several reasons. First, as the model area increases, the model calculations may become prohibitive in terms of computation time and computer memory. Second, large model areas are more difficult to parameterize, because the spatial variability of the model parameters increases with increasing model area. This almost inevitably leads to increasing uncertainty in the model outcomes, as was demonstrated by Quinton [97]. This is probably the main reason that an evaluation of different distributed field-scale and catchment-scale erosion models (including some listed in Table 3) demonstrated that the predictive quality for net soil loss of these models is 'not very good' [98]. Third, applying a model to a larger model area often involves a reduction in the spatial resolution of the model. Model parameter values derived for fine resolution models cannot simply be applied to coarse resolution models and require appropriate upscaling. As a matter of fact, this problem also arises in scale transfer from observations to models and vice versa. Many practical examples of such scale issues in environmental modelling are given by Bierkens *et al.* [99]. Finally, processes other than those represented by the model may become prevalent as the model area becomes increasingly complex with increasing size. Examples of these processes that act at scales broader than the field scale are mass movements (e.g. landslides, debris flows), sediment retention in impoundments and other buffer features (e.g. terraces, grass strips, farm ponds, hedges, roadsides, river banks), stream bank erosion, and sediment storage in channels and floodplains (see Chapter 4, this book). If the model does not (or poorly) account for these processes, this inevitably leads to additional uncertainty if the model output is incorrectly interpreted. For example, although the PESERA model (see Table 3) predicts soil losses for the whole of Europe at a spatial resolution of 1 km^2, it does not consider transport capacity-limited sediment transport (see above discussion). This implies that the model actually predicts hillslope-scale soil losses. Although this has been clearly stated by the authors [96], it may lead to erroneous overestimations of sediment transfer if these predicted soil loss rates are interpreted as soil losses from each individual grid cell. Conversely, other processes may become irrelevant at broader scales. Thus, tillage operations may considerably increase local (plot and hillslope scale) soil redistribution over most of the hilly cropland of Western Europe [100, 101]. However, tillage erosion contributes little to sediment losses at the catchment scale.

These problems related to scale are especially relevant for sediment management at the river basin scale. After all, it is not only necessary to know how much and where sediment is produced in the terrestrial part of the catchment system, but also to know how much of this sediment reaches the river network and is transported to the catchment outlet. The problems related to connectivity between hillslopes and rivers in modelling sediment transfer at the river basin scale have been extensively discussed by Beven *et al.* [102]. There are currently no adequate process-based models available that are able to predict the catchment-scale transfer of sediment eroded on hillslopes to streams. Some studies have tackled this limitation by applying the empirical concept of sediment delivery ratios (SDR), i.e. the ratio of sediment delivered at a location in the stream system to the gross erosion in the catchment area upstream from that location. For example, Asselman *et al.* [103] adopted two different sediment delivery ratios for modelling sediment transport in the River Rhine basin based on RUSLE-derived hillslope erosion estimates. The first SDR referred to the transport from hillslope to river network and the second SDR to the transfer within the river network to the catchment outlet.

For further, recent state-of-the-art reviews of modelling approaches for soil erosion and sediment redistribution in catchments see [104–106].

4.3. Sediment transport and deposition models

Although the soil erosion models presented in the previous section are able to predict sediment production in the terrestrial parts of the catchment, they generally perform less well in modelling the transport and fate of sediment in the surface water network. For this latter purpose, a variety of process-based modelling packages are available for modelling of sediment transport in river, estuarine and coastal environments (see Table 4). These models often combine results from hydrodynamic modelling with the complex physics that governs the movement of sediments. Hydrodynamic models for flow in river channels

Table 4. Examples of hydrodynamic and sediment transport models

Name	Model dimensions	Model domain	WWW
Sobek	1D	Rivers and estuaries	http://www.sobek.nl/
Delft3D	2D/3D	Rivers, estuaries and seas	http://www.wldelft.nl/soft/d3d/intro/
Mike 11	1D	Rivers and estuaries	http://www.dhigroup.com/Software/WaterResources/MIKE11.aspx
Mike 21C	2D	Rivers	http://www.dhigroup.com/Software/WaterResources/MIKE21C.aspx
SMS	1D/2D	Rivers	http://www.ems-i.com/SMS/SMS_Overview/sms_overview.html
Telemac	2D*	Rivers, estuaries and near-coastal areas	http://www.telemacsystem.com/
TRIM-2D	2D	Rivers and estuaries	http://www.baw.de/vip/en/departments/department_k/methods/hnm/trim2d/trim2-en.html
HEC6	1D	Rivers	http://www.hec.usace.army.mil/software/legacysoftware/hec6/hec6.htm
ECOMSED	1D/2D/3D	Rivers, estuaries and near-coastal areas	http://www.hydroqual.com/ehst.html
CCHE1D/ 2D/3D	1D/ 2D/3D	Rivers, estuaries and near-coastal areas	http://www.ncche.olemiss.edu/index.php?page=freesoftware
Cosmos	1D/2D/3D	Rivers, estuaries and near-coastal areas	http://www.baird.com/baird/en_html/NumModSed.html
SedNet	2D	Catchments	http://www.catchment.crc.org.au/cgi-bin/WebObjects/toolkit.woa/1/wa/productDetails?productID=1000013

* The number of spatial dimensions applies to the sediment transport module of the model; some models allow the simulation of the hydrodynamics in three dimensions.

are essentially based on the Saint-Venant equations for channel flood routing. For estuaries and coastal areas, the forces due to tidal accelerations, wind stress, wind waves, Coriolis accelerations (due to the rotation of the Earth) and buoyancy (due to density stratification) are included in the equations. All models require detailed information on the morphology of the channel or sea bed (elevations, cross-sections, hydraulic roughness). Most models (e.g. Sobek, Delft3D, CCHE, TRIM-2D) also allow the simulation of morphological changes of the bed or channel banks due to erosion and deposition processes. The models listed in Table 4 may further differ in their capabilities to distinguish between suspended and bed load transport, or to simulate the transport of multiple grain size classes.

Sediment transport models are commonly employed for predicting erosion and sedimentation rates and related morphological changes in channels, floodplains and harbours, and conveyance losses in river reaches. One-dimensional models usually suffice for modelling sediment transport through the river network. Most one-dimensional models also permit the simulation of water flow and sediment transport in multiple river and floodplain branches. For situations where lateral dispersal of sediments is important, for example in lakes, floodplains, harbours or estuaries, a two-dimensional model may be more appropriate. Obviously, two-dimensional models are more computationally demanding than one-dimensional models. Therefore, they are usually only applied to areas of limited extent (e.g. [107]). For the same reason, three-dimensional models are usually only adopted for situations where it is vital to take into account the effects of vertical stratification (e.g. in estuaries and deep lakes) or interactions between flow and bed morphology (e.g. in meander bends or bifurcations).

As most of the models presented in Table 4 are directed towards sediment fluxes in various types of surface waters, they generally do not predict sediment inputs from the terrestrial part of the river system due to soil erosion or mass movements. Some of the model packages, however, allow the integration of models of soil erosion and sediment transport in the river network, for example the Australian SedNet model.

The majority of sediment transport models utilize a Eulerian fixed grid to numerically solve the differential equations of flow and transport. A major disadvantage of this method is that it suffers from numerical dispersion, i.e. artificial mixing solely attributable to the numerical solution technique. In recent years, a number of sediment transport models have been published based on particle tracking methods, which minimize the undesired effects of numerical dispersion. Although particle tracking has been commonly accepted in modelling of conservative or degradable solute transport since the late 1970s, it had never been applied to sediment transport. Examples of particle tracking models for sediment transport are the PARTRACE post-processing module of the TRIM-2D modelling system (see Table 4) and the GIS-based MocSED model [108].

A fundamentally different approach to model suspended sediment concentrations in large river basins across the globe has been presented by Håkanson [109]. This approach is much more conceptual and empirical than the abovementioned physically based models. Håkanson's model predicts mean monthly suspended particulate matter (SPM) (including biologically produced particulates) concentrations for a river stretch based on simple, empirical process descriptions of primary production, resuspension, mineralization and retention of SPM in the upstream river stretch. The driving variables for these processes are catchment area, annual average precipitation and the location on the Earth (altitude, latitude and distance from the sea). Despite its simplicity, the model's predictive power has been demonstrated to be very good.

Another recent example of such a global-scale empirical model is the BQART model developed by Syvitski and Milliman [110]. The BQART model predicts the long-term (approximately 30 year) flux of sediment delivered by rivers to coastal waters based on simple and readily available parameters. The model accounts for geomorphological parameters (basin area, relief), climate (temperature, ice cover), hydrology (discharge), geology (lithology) and human activities (reservoir trapping, soil erosion, urbanization). Relief is defined as the altitude difference between the highest point in the drainage basin and the gauging station where observations are made. The model distinguishes six lithology types, ranging from hard, acid plutonic or high-grade metamorphic rocks to extremely weak substrates, such as crushed rock or loess deposits. Human influence is estimated using a simple, a priori method based on population density and gross national product per capita. The BQART model does not provide within-basin details on sediment erosion, transport and retention. When applied to the database of 488 test rivers, the BQART model explains 96% of the between-river variation in the long-term sediment load [110].

4.4. Water and sediment quality models

Many of the sediment transport models discussed in the previous section also have modules for predicting water quality with respect to a variety of parameters such as nutrients, dissolved oxygen, trace metals, radionuclides, pesticides and other organic pollutants. In general, these models focus on in-stream water quality processes like biological production and degradation, and sorption to suspended matter and bed sediments. More detailed information on modelling contaminant behaviour in bed sediments at the local scale is given in Allan and Stegemann [111]. Other common river water quality models include:

- WASP [112] for dynamic modelling of eutrophication processes and contaminant behaviour.
- QUAL2E [113] for modelling nutrient, algae and dissolved oxygen dynamics.

- GREAT-ER [114] for modelling the steady-state impacts of point source emissions of a variety of inorganic and organic chemicals.
- RIVTOX (1D) and COASTOX (2D) [115] for modelling transport and behaviour of radionuclides in rivers, lakes, reservoirs and estuaries.

These models are less well-suited for modelling diffuse emissions in the catchment as they only simulate in-stream processes: diffuse inputs into the river system need to be set by the user as boundary conditions. For situations where diffuse emissions are important, the pathways that represent the links between the emissions in the catchment and the inputs into the river system have to be modelled explicitly. There are a number of physically based integrated catchment models available that take account of the various pathways from the terrestrial part of the catchment to the river network, including:

- SWAT [116].
- HSPF [117].
- AGNPS [118].
- SHETRAN [119].
- TERRACE [120].

Most of these catchment-scale water quality models include modules for runoff and soil erosion. Accordingly, they require a large amount of input data, often at a detailed spatial or temporal resolution. This causes these models to be difficult to parameterize, especially for larger river basins. Furthermore, the scale issues discussed in Section 4.2 also apply to these models, perhaps even to a larger extent because these models not only take the surface pathways of erosion into account, but also the subsurface pathways of leaching and groundwater flow. That is why these water quality models are usually applied to relatively small catchments up to several square kilometres in size.

Macro-scale water and sediment quality modelling at the scale of larger river basins (>1000 km^2) requires a different modelling approach. Similarly to the abovementioned conceptual approach for predicting suspended matter in rivers [109], various conceptual approaches have been developed for the estimation of contaminant fluxes and source apportionment at the river basin scale. These approaches are based on the quantification of emissions and the resulting loads transferred via the various hydrological pathways. Models that use these approaches include HBV-N [121], MONERIS [122] and PolFlow [123]. Figure 6 shows a schematic structure of the MONERIS model; the other models have similar structures. The HBV-N and PolFlow models have hydrological modules for estimating the water flow following the different hydrological pathways; MONERIS relies on measured discharges. The retention and losses along the

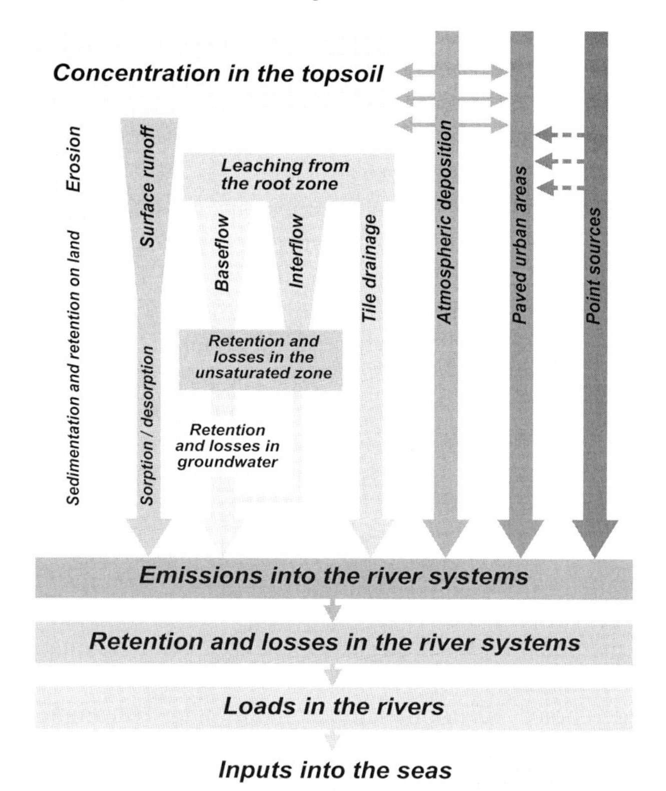

Figure 6. Pathways and processes in MONERIS (source: adapted from [125])

subsurface pathways are assumed to increase with increasing groundwater transit time, which can be estimated using aquifer properties, recharge rate and configuration of the river network. The losses in the river network are assumed to decrease with discharge (or specific runoff [124]) and slope gradient, and increase with the occurrence of lakes.

The input of these models consists of extensive yet generic and generally available geographical and statistical information (soil, land use, climate, population and livestock densities, and sewage connection rates). The PolFlow model is a GIS-embedded model and uses 1 km^2 grid cells to provide a spatially distributed representation of the river basin, whereas HBV-N and MONERIS discretize the river basin space into sub-basins. These models were originally developed and tested for modelling nutrient fluxes in various large European river basins. HBV-N has been initially applied to Scandinavian catchments and the entire Baltic Sea basin [126], MONERIS to German catchments [125], and PolFlow to the Rhine, Elbe and Po basins [123, 127]. The PolFlow model and MONERIS approaches have recently been adapted for modelling total metal fluxes in the Rhine and Elbe basins [128, 129], but they are not yet suitable for a more detailed assessment of sediment quality.

5. Summary and conclusions

There are a variety of tools available for scientists and managers to facilitate the decision-making process. The tools presented in this chapter include geographic information, census and monitoring data, tracing and fingerprinting methods, and mathematical models. Tools for site-specific risk assessment and cost-benefit analysis are discussed elsewhere in this book: for these and other tools for sediment management see also [1, 2, 10]. The tools discussed in this chapter are especially powerful in providing answers to one or more key questions in sediment management concerning sediment amounts and quality. They allow an assessment of contaminant levels, erosion and deposition rates, sediment and contaminant fluxes through river basins, and sources of sediments and contaminants. In addition, they enable the prediction of changes in erosion, sediment transport, deposition, morphology and quality as a result of changes in land use, climate and river basin management strategies. Sediment tracing and fingerprinting methods allow further assessment of the amounts and sources of sediment that is being redistributed within river basins.

Although the choice of tools and methodologies is ample, there is no integrated catchment tool available which assists in all aspects and phases of the sequential process of decision-making on sediment management at the river basin scale. Consequently, decision-making must largely rely on models and methods that focus on separate aspects and steps in the decision-making process. Some of these tools are barely beyond the stage of research tools and some are still under development, though they have great potential to aid in decision-making. Hence, there is a need to develop better and more integrated catchment-scale models of sediment dynamics addressing the complexities inherent in managing sediments at scales ranging from contaminated sites to entire river basins. The current GIS-based modelling approaches allow for this kind of integration. However, such approaches have been primarily adopted for modelling water quality and contaminant fluxes in river basins. Although several attempts have been made to model basin-scale sediment transport, successful predictions are often hampered by our limited knowledge on sediment connectivity at hillslope-stream channel transitions. Moreover, these models do not, or only poorly, account for temporal storage effects in the channel bed, floodplains, lakes and wetlands. Therefore, existing approaches of macro-scale modelling should be complemented and modified to incorporate all relevant processes for the estimation of sediment transport and contamination at the river basin scale. For a successful application of basin-scale sediment transport models, they should effectively account for sediment connectivity and temporal storage effects in the channel bed, floodplains, lakes and wetlands. Furthermore, there is a need for models which couple better sediment fluxes and dynamics in the riverine parts of 'river basins' with those in the estuarine and coastal parts of the basin.

References

1. Heise S (Ed.) (2007). Sustainable Management of Sediment Resources: Sediment Risk Management and Communication. Elsevier, Amsterdam.
2. Bortone G, Palumbo L (Eds) (2007). Sustainable Management of Sediment Resources: Sediment and Dredged Material Treatment. Elsevier, Amsterdam.
3. Isaaks EH, Srivastava RM (1989). Applied Geostatistics. Oxford University Press, New York.
4. Wierzbicki A, Makowski M, Wessels J (Eds) (2000). Model-based Decision Support Methodology with Environmental Applications. Kluwer, Dordrecht.
5. Ehrgott M (2005). Multicriteria Optimization, 2nd edn. Springer, Berlin.
6. Foster IDL, Charlesworth SM, Keen DH (1991). A comparative study of heavy metal contamination and pollution in four reservoirs in the English Midlands, UK. Hydrobiologia, 214: 155–162.
7. Valette-Silver NJ (1993). The use of sediment cores to reconstruct historical trends in contamination of estuarine and coastal sediments. Estuaries, 16: 577–588.
8. Middelkoop H (2000). Heavy-metal pollution of the River Rhine and Meuse floodplains in the Netherlands. Geologie en Mijnbouw/Netherlands Journal of Geosciences, 79: 411–428.
9. Gallagher L, Macdonald RW, Paton DW (2004). The historical record of metals in sediments from six lakes in the Fraser River Basin, British Columbia. Water, Air and Soil Pollution, 152: 257–278.
10. Barceló D, Petrovic M (Eds) (2007). Sustainable Management of Sediment Resources: Sediment Quality and Impact Assessment of Pollutants. Elsevier, Amsterdam.
11. Burrough PA, McDonnell RA (1998). Principles of Geographical Information Systems. Oxford University Press, Oxford.
12. Timmerman J, Langaas S (Eds) (2003). Environmental Information in European Transboundary Water Management. IWA Publishing, London.
13. Heise S, Förstner U, Westrich B, Salomons W, Schönberger H (2004). Inventory of historical contaminated sediment in Rhine Basin and its tributaries – Final report. Technical University Hamburg Harburg, Hamburg.
14. Meybeck M, Kimstach V, Helmer R (1992). Strategies for water quality assessment. In: Chapman D (Ed.), Water Quality Assessments – A Guide to Use of Biota, Sediments and Water in Environmental Monitoring, 2nd edn. E&FN Spon, London.
15. Rode M, Suhr U (2006). Uncertainties in selected surface water quality data. Hydrology and Earth Systems Science, Discussion, 3: 2991–3021.
16. DK Rhein (2006). Zahlentafeln der chemisch-physikalischen Untersuchungen 1999, 2000, 2001, 2002, and 2003. Deutsche Kommission zur Reinhaltung des Rheins, Worms, Germany. Available online at http://www.dk-rhein.de/ [verified 22 October 2006].
17. Rijkswaterstaat (2006). Waterstat: Aggregated Measurement Data of the Ministry of Transport, Public Works, and Water Management, the Netherlands. Available online at http://www.waterstat.nl/ [verified 22 October 2006].
18. Van der Perk M (2006). Soil and Water Contamination: From Molecular to Catchment Scale. Taylor & Francis/Balkema, Leiden.

19. Cohn TA (1995). Recent advances in statistical methods for the estimation of sediment and nutrient transport in rivers. Reviews in Geophysics, 33: 1117–1124.
20. Quilbé R, Rousseau AN, Duchemin M, Poulin A, Gangbazo G, Villeneuve J-P (2006). Selecting a calculation method to estimate sediment and nutrient loads in streams: application to the Beaurivage River (Québec, Canada). Journal of Hydrology, 326: 295–310.
21. Walling DE (1977). Assessing the accuracy of suspended sediment rating curves for a small basin. Water Resources Research, 13: 531–538.
22. Ferguson RI (1986). River loads underestimated by rating curves. Water Resources Research, 22: 74–76.
23. Horowitz A (2003). An evaluation of sediment rating curves for estimating suspended sediment concentrations for subsequent flux calculations. Hydrological Processes, 17: 3387–3409.
24. Asselman NEM (2000). Fitting and interpretation of sediment rating curves. Journal of Hydrology, 234: 228–248.
25. Middelkoop H, Thonon I, Van der Perk M (2002). Effective discharge for heavy metal deposition on the lower River Rhine flood plains. In: Dyer FJ, Thoms MC, Olley JM (Eds), The Structure, Function and Management Implications of Fluvial Sedimentary Systems. IAHS Publication no. 276. IAHS Press, Wallingford, pp. 151–160.
26. Gandrass J, Salomons W (Eds) (2001). Dredged Material in the Port of Rotterdam – Interface between Rhine Catchment Area and North Sea. Present and Future Quality of Sediments in the Rhine Catchment Area. GKSS Research Centre, Institute for Coastal Research, Geesthacht, Germany.
27. Grimvall A, Stålnacke P (1996). Statistical methods for source apportionment of riverine loads of pollutants. Environmetrics, 7: 201–213.
28. Rogowski AS, Tamura T (1965). Movement of Cs-137 by runoff, erosion and infiltration on the alluvial Capitina silt loam. Health Physics, 11: 1333–1340.
29. Ritchie JC, Spraberry JA, McHenry JR (1974). Estimating soil erosion from the redistribution of fallout Cs-137. Soil Science Society of America Proceedings, 38: 137–139.
30. Campbell BL, Loughran RJ, Elliot GL (1982). Caesium-137 as an indicator of geomorphic processes in a drainage basin system. Australian Geographical Studies, 20: 49–64.
31. Wall GJ, Wilding LP (1976). Mineralogy and related parameters of fluvial suspended sediments in northwestern Ohio. Journal of Environmental Quality, 5: 168–173.
32. Walling DE, Peart MR, Oldfield F, Thompson R (1979). Suspended sediment sources identified by magnetic measurements. Nature, 281: 110–113.
33. Peart MR, Walling DE (1988). Techniques for establishing suspended sediment sources in two drainage basins in Devon, UK: a comparative assessment. In: Bordas MP, Walling DE (Eds), Sediment Budgets. IAHS Publication no. 174. IAHS Press, Wallingford, pp. 269–279.
34. Walling DE (2006). Tracing versus monitoring: new challenges and opportunities in erosion and sediment delivery research. In: Owens PN, Collins AJ (Eds), Soil Erosion and Sediment Redistribution in River Catchments: Measurement, Modelling and Management. CABI Publishing, Wallingford, pp. 13–27.
35. Walling DE (2004). Using environmental radionuclides to trace sediment mobilisation and delivery in river basins as an aid to catchment management. Proceedings of the Ninth International Symposium on River Sedimentation, 18–21 October 2004, Yichang, China, pp. 121–125.
36. Walling DE (2005). Tracing suspended sediment sources in catchments and river systems. The Science of the Total Environment, 344: 159–184.

37. Reid LM, Dunne T (1996). Rapid Evaluation of Sediment Budgets. GeoEcology Paperbacks, Catena Verlag, Germany.
38. Wallbrink PJ, Walling DE, He Q (2002). Radionuclide measurement using HPGe gamma spectrometry. In: Zapata F (Ed.), Handbook for the Assessment of Soil Erosion and Sedimentation Using Environmental Radionuclides. Kluwer, Dordrecht, pp. 67–96.
39. Owens PN, Walling DE (1996). Spatial variability of caesium-137 inventories at reference sites: an example from two contrasting sites in England and Zimbabwe. Applied Radiation and Isotopes, 47: 699–707.
40. Robbins JA (1978). Geochemical and geophysical applications of radioactive lead. In: Nrigu JO (Ed.), Biochemistry of Lead in the Environment. Elsevier Scientific, Amsterdam, pp. 285–393.
41. Wallbrink PJ, Murray AS (1994). Fallout of ^{7}Be in South Eastern Australia. Journal of Environmental Radioactivity, 25: 213–228.
42. Ritchie JC, McHenry JR (1990). Application of radioactive fallout cesium-137 for measuring soil erosion and sediment accumulation rates and patterns: a review. Journal of Environmental Quality, 19: 215–233.
43. Walling DE, Quine TA (1990). Calibration of caesium-137 measurements to provide quantitative erosion rate data. Land Degradation and Rehabilitation, 2: 161–175.
44. Walling DE, He Q (1999). Improved models for estimating soil erosion rates from ^{137}Cs measurements. Journal of Environmental Quality, 28: 611–622.
45. Walling DE, He Q, Appleby PG (2002). Conversion models for use in soil-erosion, soil-redistribution and sedimentation investigations. In: Zapata F (Ed.), Handbook for the Assessment of Soil Erosion and Sedimentation Using Environmental Radionuclides. Kluwer, Dordrecht, pp. 67–96.
46. Blake WH, Walling DE, He Q (1999). Fallout ^{7}Be as a tracer in soil erosion investigations. Applied Radiation and Isotopes, 51: 599–605.
47. Wilson CG, Matisoff G, Whiting PJ (2003). Short-term erosion rates from a Be-7 inventory balance. Earth Surface Processes and Landforms, 28: 967–977.
48. Ritchie JC, Ritchie CA (2007). Bibliography of publications of ^{137}Cs studies related to soil erosion and sediment redistribution. Available online at http://hydrolab.arsusda.gov/cesium/ [verified 25 April 2007].
49. Walling DE, He Q, Blake WH (1999). Use of ^{7}Be and ^{137}Cs measurements to document short- and medium-term rates of water-induced soil erosion on agricultural land. Water Resources Research, 35: 3865–3874.
50. Wallbrink PJ, Roddy BP, Olley JM (2002). A tracer budget quantifying soil redistribution on hillslopes after forest harvesting. Catena, 47: 179–201.
51. Wallbrink PJ, Murray AS, Olley JM, Olive LJ (1998). Determining sources and transit times of suspended sediment in the Murrumbidgee River, New South Wales, Australia, using fallout Cs-137 and Pb-210. Water Resources Research, 34: 879–887.
52. He Q, Walling DE, Owens PN (1996). Interpreting the Cs-137 profiles observed in several small lakes and reservoirs in southern England. Chemical Geology, 129: 115–131.
53. Appleby PG (2001). Chronostratigraphic techniques in recent sediments. In: Last WM, Smol JP (Eds), Tracking Environmental Change Using Lake Sediments. Vol. 1: Basin Analysis, Coring and Chronological Techniques. Kluwer Academic, Dordrecht, pp. 171–203.
54. Wallbrink PJ, Murray AS, Olley JM (1999). Relating suspended sediment to its original soil depth using fallout radionuclides. Soil Science Society of America Journal, 63: 369–378.
55. Collins AL, Walling DE (2004). Documenting catchment suspended sediment sources: problems, approaches and prospects. Progress in Physical Geography, 28: 159–196.

56. Horowitz AJ (1991). A Primer on Sediment Trace-Element Chemistry. Lewis Publishers, Michigan.
57. Wallbrink PJ (2004). Quantifying the erosion processes and land-uses which dominate the supply of sediment to Moreton Bay, S.E. Queensland, Australia. Journal of Environmental Radioactivity, 76: 67–80.
58. Parker A, Bergmann H, Heininger P, Leeks GJL, Old GH (2007). Sampling of sediments and suspended matter. In: Barceló D, Petrovic M (Eds), Sustainable Management of Sediment Resources: Sediment Quality and Impact Assessment of Pollutants. Elsevier, Amsterdam, pp. 1–34.
59. Carter J, Owens PN, Walling DE, Leeks GJL (2003). Fingerprinting suspended sediment sources in a large urban river system. The Science of the Total Environment, 314–316: 513–534.
60. Collins AL, Walling DE, Leeks GJL (1997). Fingerprinting the origin of fluvial suspended sediment in larger river basins: combining assessment of spatial provenance and source type. Geografiska Annaler, 79A: 239–254.
61. Klages MG, Hsieh YP (1975). Suspended solids carried by the Gallatin River of Southwestern Montana: II. Using mineralogy for inferring sources. Journal of Environmental Quality, 4: 68–73.
62. Yu L, Oldfield F (1989). A multivariate mixing model for identifying sediment sources from magnetic measurements. Quaternary Research, 32: 168–181.
63. Walden J, Slattery MC, Burt TP (1997). Use of mineral magnetic measurements to fingerprint suspended sediment sources: approaches and techniques for data analysis. Journal of Hydrology, 202: 353–372.
64. Walden J, Oldfield F, Smith J (1999). Environmental Magnetism: A Practical Guide. Technical Guide 6. Quaternary Research Association, Cambridge/London.
65. Granger SJ, Bol R, Butler PJ, Naden P, Old G, Owens PN, Smith B (2007). Processes affecting transfer of sediment and colloids, with associated phosphorus, from intensively farmed grasslands: tracing sediment and organic matter. Hydrological Processes, 21: 417–422.
66. Bricelj M, Misic M (1997). Movement of bacteriophage and fluorescent tracers through underground river sediments. In: Kranjc A (Ed.), Tracer Hydrology. Balkema, Rotterdam, pp. 3–9.
67. Marsh JK, Bale AJ, Uncles RJ, Dyer KR (1991). A tracer technique for the study of suspended sediment dynamics in aquatic environments. IAHR International Symposium on the Transport of Suspended Sediments and its Mathematical Modelling, Florence, Italy.
68. Young RA, Holt RF (1968). Tracing soil movement with fluorescent glass particles. Soil Science Society of America Proceedings, 32: 600–602.
69. Ventura EJ, Nearing MA, Norton LD (2001). Developing a magnetic tracer to study soil erosion. Catena, 43: 277–291.
70. Polyakov VO, Nearing MA (2004). Rare earth element oxides for tracing sediment movement. Catena, 55: 255–276.
71. Zhang XC, Friedrich JM, Nearing MA, Norton LD (2001). Potential use of rare earth oxides as tracers for soil erosion and aggregation studies. Soil Science Society of America Journal, 65: 1508–1515.
72. Mahler BJ, Winkler M, Bennett P, Hillis DM (1998). DNA-labelled clay: a sensitive new method for tracing particle transport. Geology, 26: 831–834.
73. Götz R, Steiner B, Friesel P, Roch K, Walkow F, Maaß V, Reincke H, Stachel B (1998). Dioxin (PCDD/F) in the River Elbe – investigations of their origin by multivariate statistical methods. Chemosphere, 37: 1987–2002.
74. McConnachie JL, Petticrew EL (2006). Tracing organic matter sources in riverine suspended sediment: implications for fine sediment transfers. Geomorphology, 79: 13–26.

75. Collins AL, Walling DE, Leeks GJL (1998). The use of composite fingerprints to determine the provenance of the contemporary suspended sediment load transported by rivers. Earth Surface Processes and Landforms, 23: 31–52.
76. Rowan JS, Goodwill P, Franks SW (2000). Uncertainty estimation in fingerprinting suspended sediment sources. In: Foster IDL (Ed.), Tracers in Geomorphology. Wiley, Chichester, pp. 279–290.
77. Motha JA, Wallbrink PJ, Hairsine PB, Grayson RB (2002). Tracer properties of eroded sediment and source material. Hydrological Processes, 16: 1983–2000.
78. Foster IDL (2006). Lakes and reservoirs in the sediment delivery system. In: Owens PN, Collins AJ (Eds), Soil Erosion and Sediment Redistribution in River Catchments: Measurement, Modelling and Management. CABI Publishing, Wallingford, pp. 13–27.
79. Evans DJ, Gibson CE, Rossell RS (2006). Sediment loads and sources in heavily modified Irish catchments: a move towards informed management strategies. Geomorphology, 79: 93–113.
80. Sear DA, Lee MWE, Oakey RJ, Carling PA, Collins MB (2000). Coarse sediment tracing technology in littoral and fluvial environments: a review. In: Foster IDL (Ed.), Tracers in Geomorphology. Wiley, Chichester, pp. 21–55.
81. Custer SG, Ergenzinger P, Anderson B, Bugosh N (1987). Electromagnetic detection of pebble transport in streams: a method for measurement of sediment-transport waves. In: Ethridge FG, Flores R (Eds), Recent Developments in Fluvial Sedimentology. Society of Economic Paleontologists and Mineralogists, Special Publication 39, pp. 21–26.
82. Haschenburger JK, Church M (1998). Bed material transport estimated from the virtual velocity of sediment. Earth Surface Processes and Landforms, 23: 791–808.
83. Busskamp R, Hasholt B (1996). Coarse bedload transport in a glacial valley, Sermilik, southeast Greenland. Zeitschrift für Geomorphologie, 40: 349–358.
84. Morgan RPC, Quinton JN (2001). Erosion modeling. In: Harmon RS, Doe III WW (Eds), Landscape Erosion and Evolution Modeling. Kluwer Academic/Plenum, New York, pp. 117–143.
85. Harmon RS, Doe III WW (Eds) (2001). Landscape Erosion and Evolution Modeling. Kluwer Academic/Plenum, New York.
86. Summer W, Walling DE (Eds) (2001). Modelling erosion, sediment transport and sediment yield. Technical Documents in Hydrology No. 60. UNESCO, Paris.
87. Morgan RPC, Quinton JN, Smith RE, Govers G, Poesen JWA, Auerswald K, Chisci G, Torri D, Styczen ME (1998). The European soil erosion model (EUROSEM): a process-based approach for predicting sediment transport from fields and small catchments. Earth Surface Processes and Landforms, 23: 527–544.
88. De Roo APJ, Wesseling CG, Ritsema CJ (1996). LISEM: a single-event physically based hydrological and soil erosion model for drainage basins. I: Theory, input and output. Hydrological Processes, 10: 1107–1117.
89. Bouraoui F, Dillaha TA (1996). ANSWERS-2000: Runoff and sediment transport model. Journal of Association of American Engineers, ASCE, 122: 493–502.
90. Laflen JM, Lane LJ, Foster GR (1991). WEPP, a new generation of erosion prediction technology. Journal of Soil and Water Conservation, 46: 34–38.
91. Wischmeier WH, Smith DD (1978). Predicting rainfall-erosion losses: a guide to conservation farming. Agricultural Handbook No. 537. US Dept of Agriculture, Washington DC.
92. Renard KG, Foster GR, Weesies GA, Porter JP (1991). RUSLE: revised universal soil loss equation. Journal of Soil and Water Conservation, 46: 30–33.
93. Leonard RA, Knisel WG, Still DA (1987). GLEAMS: groundwater loading effects of agricultural management systems. Transactions of the American Society of Agricultural Engineers, 30: 1403–1418.

94. Morgan RPC (2001). A simple approach to soil loss prediction: a revised Morgan–Morgan–Finney model. Catena, 44: 305–322.
95. Van Rompaey A, Verstraeten G, Van Oost K, Govers G, Poesen J (2001). Modelling mean annual sediment yield using a distributed approach. Earth Surface Processes and Landforms, 26: 1221–1236.
96. Govers G, Gobin A, Cerdan O, van Rompaey A, Kirkby M, Irvine B, Le Bissonais Y, Daroussin J, King D, Jones RJA, Montanarella L, Grimm M, Vieillefont V, Puigdefabregas J, Boer M, Yassoglou N, Kosmas C, Tsara M, van Lynden G, Mantel S (2003). Pan-European soil erosion risk assessment for Europe: the PESERA Map. JRC, Ispra, Italy. Available online at http://eusoils.jrc.it/ESDB_Archive/pesera/pesera_cd/pdf/ThePeseraMap.pdf [verified 25 October 2006].
97. Quinton JN (1997). Reducing predictive uncertainty in model simulations: a comparison of two methods using the European Soil Erosion Model. Catena, 30: 101–117.
98. Jetten V, De Roo A, Favis-Mortlock D (1999). Evaluation of field-scale and catchment-scale soil erosion models. Catena, 37: 521–541.
99. Bierkens MFP, Finke PA, De Willigen P (2000). Upscaling and Downscaling Methods for Environmental Research. Kluwer Academic, Dordrecht.
100. Govers G, Vandaele K, Desmet P, Poesen J, Bunte K (1994). The role of tillage in soil redistribution on hillslopes. European Journal of Soil Science, 45: 469–478.
101. Van Oost K, Govers G, De Alba S, Quine TA (2006). Tillage erosion: a review of controlling factors and implications for soil quality. Progress in Physical Geography, 30: 443–466.
102. Beven KJ, Heathwaite AL, Haygarth P, Walling D, Brazier RE, Withers P (2005). On the concept of delivery of sediment and nutrients to stream channels. Hydrological Processes, 19: 551–556.
103. Asselman NEM, Middelkoop H, Van Dijk PM (2003). The impact of climate change on soil erosion, transport and deposition of suspended sediment in the River Rhine. Hydrological Processes, 17: 3225–3244.
104. Merritt WS, Letcher RA, Jakeman AJ (2003). A review of erosion and sediment transport models. Environmental Modelling and Software, 18: 761–799.
105. Brazier RE (2004). Quantifying soil erosion by water in the UK: a review of monitoring and modeling approaches. Progress in Physical Geography, 28: 340–365.
106. Owens PN, Collins AJ (Eds) (2006). Soil Erosion and Sediment Redistribution in River Catchments: Measurement, Modelling and Management. CABI, Wallingford.
107. Middelkoop H, Van der Perk M (1998). Modelling spatial patterns of overbank sedimentation on embanked floodplains. Geografiska Annaler, 80A: 95–109.
108. Thonon I, De Jong K, Van der Perk M, Middelkoop H (2007). Modelling floodplain sedimentation using particle tracking. Hydrological Processes, 21: 1402–1412.
109. Håkanson L (2006). A dynamic model for suspended particulate matter (SPM) in rivers. Global Ecology and Biogeography, 15: 93–107.
110. Syvitski JPM, Milliman JD (2007). Geology, geography, and humans battle for dominance over the delivery of sediment to the coastal ocean. Journal of Geology, 115: 1–19.
111. Allan IJ, Stegemann JA (2007). Modelling of pollutant fate and behaviour in bed sediments. In: Barceló D, Petrovic M (Eds), Sustainable Management of Sediment Resources: Sediment Quality and Impact Assessment of Pollutants. Elsevier, Amsterdam, pp. 263–294.
112. Ambrose RB, Wool TA, Martin JL (1993). The Water Quality Analysis Simulation Program WASP5, Part A: Model Documentation, Version 5.10. US Environmental Protection Agency, Environmental Research Laboratory, Athens, GA.
113. Barnwell TO, Brown LC (1987). The Enhanced Stream Water Quality Models QUAL2E and QUAL2E-UNCAS: Documentation and User Manual, EPA/600/3-87/007.

114. Feijtel T, Boeije G, Matthies M, Young A, Morris G, Gandolfi C, Hansen B, Fox K, Matthijs E, Koch V, Schroder R, Cassani G, Schowanek D, Rosenblom J, Holt M (1998). Development of a geography-referenced regional exposure assessment tool for European rivers – GREAT-ER. Journal of Hazardous Materials, 61: 59–65.
115. Zheleznyak MJ (1997). The mathematical modelling of radionuclide transport by surface water flow from the vicinity of the Chornobyl nuclear power plant. Condensed Matter Physics, 12: 37–50.
116. Neitsch SL, Arnold JG, Kiniry JR, Williams JR (2001). Soil and Water Assessment Tool – Theoretical Documentation – Version 2000. Blackland Research Center, Agricultural Research Service, Texas.
117. Bicknell BR, Imhoff JC, Kittle JL Jr, Donigian AS Jr, Johanson RC (1997). Hydrological Simulation Program–Fortran, User's manual for version 11. US Environmental Protection Agency, National Exposure Research Laboratory, Athens, GA, EPA/600/R-97/080.
118. Bingner RL, Theurer FD, Cronshey RG, Darden RW (2001). AGNPS Web Site. Available at http://www.ars.usda.gov/Research/docs.htm?docid=5199 [verified 30 October 2006].
119. Ewen J, Parkin G, O'Connell PE (2000). SHETRAN: distributed river basin flow and transport modeling system. Journal of Hydraulic Engineering, 5: 250–258.
120. White S (2003). TERRACE – Terrestrial runoff modelling for risk assessment of chemical exposure. Geophysical Research Abstracts, 5: 09688.
121. Petterson A, Arheimer B, Johansson B (2001). Nitrogen concentrations simulated with HBV-N: new response function and calibration strategy. Nordic Hydrology, 32: 227–248.
122. Behrendt H, Bachor A (1998). Point and diffuse load of nutrients to the Baltic Sea by river basins of North East Germany (Mecklenburg-Vorpommern). Water Science and Technology, 38: 147–155.
123. De Wit MJM (2001). Nutrients fluxes at the river basin scale. I: The PolFlow model. Hydrological Processes, 15: 743–759.
124. Behrendt H, Opitz D (1999). Retention of nutrients in river systems: dependence on specific runoff and hydraulic load. Hydrobiologia, 410: 111–122.
125. Behrendt H, Huber P, Kornmilch M, Opitz D, Schmoll O, Scholz G, Uebe R (2000). Nutrient Emissions into River Basins of Germany. UBA-Texte 23/00: 1-266. Federal Environmental Agency, Berlin.
126. Petterson A, Brandt M, Lindström G (2000). Application of the HBV-N model to the Baltic Sea Drainage Basin. Vatten, 56: 7–13.
127. De Wit M, Bendoricchio G (2001). Nutrient fluxes in the Po basin. The Science of the Total Environment, 273: 147–161.
128. Vink RJ (2002). Heavy metal fluxes in the Elbe and Rhine river basins: analysis and modelling. PhD thesis, Vrije Universiteit, Amsterdam.
129. Vink R, Peters S (2003). Modelling point and diffuse heavy metal emissions and loads in the Elbe basin. Hydrological Processes, 17: 1307–1328.

Costs and Benefits of Sediment Management

A.F.L. Slob[a], J. Eenhoorn[b], G.J. Ellen[a], C.M. Gómez[c], J. Kind[d] and J. van der Vlies[a]

[a]*TNO Environment and Geosciences, Postbus 49, 2600 AA Delft, The Netherlands*
[b]*Rijkswaterstaat Bouwdienst, Postbus 20000, 3502 LA Utrecht, The Netherlands*
[c]*Universidad de Alcalá de Henares, Facultad de Ciencias Económicas, Plaza de la Victoria s/n, 288802 Alcalá de Henares, Spain*
[d]*Rijkswaterstaat RIZA, Postbus 17, 8200 AA Lelystad, The Netherlands*

1. Introduction

Different 'options' for sediment management will usually be available for a given situation. In economics, several instruments and tools have been developed to recognize and evaluate these options in a rational way. This chapter is about economical and financial tools to support the decision-making process, which forms part of a much broader sediment management framework (see Chapter 2, this book); other tools, such as monitoring, modelling and tracing techniques, are described in Chapter 5 (this book). A modern instrument is Societal Cost-Benefit Analysis (SCBA). In the literature 'Societal Cost-Benefit Analysis' and 'Social Cost-Benefit Analysis' are used indifferently. We will use here the term 'societal', because this reflects the purpose of SCBA: underpinning decisions that are beneficial to society. Decision-makers can be supported in a well-balanced way by evaluating the different options with the help of a SCBA. In this chapter we will describe the development of SCBA (Section 2) and the way it can be applied to sediment management (Section 3). Then we will present two examples of application of economic analyses for sediment and water management. The first example describes a SCBA that was applied to the dredging of sediments in the Netherlands (Section 4). Another example is an economic analysis with respect to river basin management (Section 5). This example is included here because the EU Water Framework Directive is an important driver for this type of analysis and it is a good illustration of the direction that the application of economic instruments for sediments may take. The subject of liability around sediment issues is touched upon in Section 6. The liability issue may become a major lever to raise awareness about, amongst other things, sediment issues, which can provoke an accelerated attention for policy measures with respect to (contaminated) sediment issues. The last section gives a summary of the chapter.

2. Societal Cost-Benefit Analysis

Everyone is used to the rationality of making decisions on the basis of a balance of gains and losses, or advantages and disadvantages. The idea behind such a balancing approach is that we only do things that yield us net gains and, when we can choose between alternatives, we choose the one that offers us the greatest net gain. This is the simple foundation of cost-benefit analysis. However, cost-benefit analysis (CBA) defines costs and benefits in a particular way, and it stretches the idea of an individual's balancing of costs and benefits to society's balancing of costs and benefits [1]. Behind CBA lies the paradigm of strict rationality: every actor is acting in such a way that net gains are generated. This approach can be elegant, but also has problems. The question is whether all actors act in a strictly rational way. Furthermore, CBA must often deal with effects that affect well-being (e.g. a decreased feeling of safety or loss of nature) that cannot be expressed in a straightforward way in terms of money, as is the case with the so-called welfare effects, e.g. a rise in income. These so called 'imponderables' and 'intangibles' have to be valued and balanced against the other effects, which may cause debates because the appreciation of these effects may vary widely between actors and situations.

Costs and benefits are defined according to the satisfaction of needs or preferences. Formally, everything is a benefit that increases human/societal well-being, and everything is a cost that reduces human/societal well-being. For the economist, whether or not well-being may be affected is to be discovered by looking at people's preferences. If an individual states a preference for situation A, then the benefits of moving to A must be positive for that individual. Why A is preferred is not the immediate concern, although no one would argue that the individual should not be allowed to get to situation A if it involves some immoral or illegal act. This is subject to the wider considerations about the 'morality' of allowing people 'to get whatever they want'. CBA functions on the basis that a 'better' allocation of resources should meet people's preferences [1].

The example above concerns the individual, but what is required when more people are affected by a certain decision? The instrument of Societal CBA is developed for this type of question. The word 'Societal' is used in the literature to refer to three different aspects of a CBA. First, it denotes the idea that in the evaluation the effect of the project on *all* individuals in society is included, not only on the parties directly involved (consumers and producers of the project). Second, it recognizes that distributional effects are being included. Without the distributional effects one is making an economical rather than a societal evaluation. Third, SCBA is used in situations where markets are imperfect and market prices are not always reflecting the individual's willingness to pay. A societal price would therefore mean that the market price should include effects that the market does not record, or records imperfectly. The word 'societal' is used to stress that one is attempting to give full expression to the preferences of all individuals, whether they are rich or poor, or directly or indirectly affected by the project [2].

The broad purpose of SCBA is to support decision-making that is beneficial to society. More specifically, the objective is to facilitate the more efficient allocation of society's resources. There are two major types of SCBA and two subtypes of SCBA. First of all is *ex-ante* SCBA, and this type assists with the decision about whether scarce societal resources should be allocated by government to a specific policy – whether a programme, project or piece of regulation. Thus its contribution to public policy decision-making is direct, immediate and specific. The second type is *ex post* analysis and is conducted at the end of a project. At the end, all of the costs are 'sunk' in the sense that they measure what choices have been made for the project. There is also less uncertainty about what the actual benefits and costs are. The value of such analyses is broader and less immediate as they provide information not only about the particular intervention but also about the 'class' of such interventions. In other words, such analyses contribute to 'learning' by government managers, politicians and academics about whether particular classes or types of projects are worthwhile. Eventually the weight of evidence may lead to a policy change. The first sub-type is a SCBA that is performed during the course of the life of a project and is called *in medias res*. Some elements of such studies are similar to an *ex ante* analysis, while others are similar to an *ex post* analysis. The final type of SCBA compares *ex ante* predictions with *ex post* measurements or, more likely, with *in medias res* estimates for the same project. This *comparative* type of SCBA is most useful to policy-makers to learn about the effectiveness of SCBA as a decision-making and evaluative tool [3]. Table 1 summarizes the ways in which the various types of SCBA serve different purposes. There are different approaches to performing a SCBA [2–4]. In Table 2 we present the nine basic steps of SCBA as described by Boardman *et al.* [3].

As with many methodologies and theories, CBA has been widely discussed among scientists. According to Self [5] and Lohmann [6], an important 'defect' in the CBA theory is that cost-benefit analysts claim that it is an objective technique or yardstick for recommending a policy decision. 'They are claiming in the first place that it is possible to quantify in monetary terms all sorts of

Table 1. Different types of Societal Cost-Benefit Analysis (source: modified from [3])

Value	*Ex ante*	*In medias res*	*Ex post*	*Comparative*
Resource allocation decision for this project	Yes – helps to select the best project or make 'go' vs 'no-go' decisions, if accurate	If low 'sunk' costs, can still shift resources. If high sunk costs, usually recommends continuation	Too late – the project is over	Same as *in medias res* or *ex post* analysis
Learning about actual value of specific project	Poor estimate – high uncertainty about future benefits and costs	Better – reduced uncertainty	Excellent – although some errors may remain. May have to wait a long time for this information	Same as *in medias res* or *ex post* analysis
Contributing to learning about actual value of similar projects	Unlikely to add much	Good – although contribution increases as SCBA is performed later. Need to adjust for uniqueness	Very useful – although some errors remain. Need to adjust for uniqueness. May have to wait a long time for this information	Same as *in medias res* or *ex post* analysis
Learning about omission forecasting, measurement and evaluation errors in SCBA	No	No	No	Yes – provides information about these errors and about the accuracy of SCBA for similar projects

factors that normally are not so expressed, secondly that the money terms used in the analysis really do possess the common property which they appear to have, and thirdly that these figures represent measurements of some concept of community welfare which can or should stand, if not as a unique criterion for decision-makers, then at least as one important criterion of the best policy' [5].

Table 2. The realities of doing a Societal Cost-Benefit Analysis (source: modified from [3])

The theoretical steps of a SCBA	The reality of doing a SCBA
Decide whose benefits and costs count	Contentious whether global, national, regional or local perspective is appropriate
Select the portfolio of alternative projects	Potentially infinite, the analyst should select an appropriate subset
Make an inventory of potential (physical) impacts and select measurement indicators	Difficult to identify specific impacts where unresearched scientific or biological processes are involved. True impacts may be unobservable
Predict quantitative impacts over the life of the project	Prediction is difficult, especially over long periods for complex systems
Monetize (attach Dollar or Euro values to) all impacts	Sometimes appropriate market values don't exist. Often the most important benefits are the most difficult to measure
Discount for time to find present values for costs or benefits arising over extended periods (years)	Different theories suggest different societal discount rates
Sum: add all benefits and costs (separately)	Some argument about the appropriate decision criterion
Perform a sensitivity analysis	Potentially infinite, the analyst must select an appropriate subset
Recommend the alternative with the largest net societal benefits	This is usually easy. It normally does not present any practical analytical difficulties, just political ones. The one exception is where sensitivity analysis shows that net present value estimates are very uncertain

The main point of criticism is that it is a tool based on an 'excess of rationality' and tries to rationalize what cannot be rationalized. Decisions that may have effects like the destruction of nature, damage to health (both animals and human beings) or even loss of life, for example, are especially difficult to rationalize because these effects are difficult to value.

But the tool of CBA is nevertheless used. This is often because of a practical vision of CBA as clearly described by Kelman: 'Nonetheless, we do not dispute that cost-benefit analysis is highly imperfect. We would welcome a better guide to public policy, a guide that would be efficient, morally attractive, and certain to ensure that governments follow the dictates of the governed. However, the decisions that must be made by contemporary decision makers do involve painful choices. They affect both the absolute quantity and the distribution of not only goods and benefits but also of physical and mental suffering. It is easy to understand why people would want to avoid making such choices and would rather act in ignorance than with knowledge and responsibility for the consequences of their choices. While this may be understandable, I do not regard it as an acceptable moral position. To govern is to choose, and decision makers – whether elected or appointed – betray their obligations to the welfare of the people who hired them if they adopt a policy of happy ignorance and non responsibility for consequences' [7].

3. Sediment management and Societal Cost-Benefit Analysis

Societal Cost-Benefit Analysis has been often applied to water management issues [8, 9], whereas its application to sediment issues is not very widespread. As was explained earlier, it is a tool with a long history that, at least potentially, contributes to a better, transparent decision-making processes. However, this implies a good understanding of the relevant system, the definition of the problems and the specification of the alternative solutions to these problems. In modern SCBA applications, this also implies the involvement of stakeholders (see [9] and Chapter 7, this book), not only to get insight into the system but also to specify the alternative actions that should be evaluated in the SCBA. A method that can help to streamline this process is Joint Fact-Finding (JFF). JFF can help the parties involved to resolve factual disagreements in ways that are acceptable to all parties. In JFF, stakeholders with differing viewpoints and interests work together to develop data and information, analyse facts and forecasts, develop common assumptions and informed opinions and, finally, use the information they have developed to reach decisions together [10].

A very crucial condition in the application of SCBA for sediment management is, therefore, to define the policy problem and the alternative actions to solve this problem. This may sound trivial, but is not as easy as may seem at first glance. One must bear in mind that economists or experts in the field of sediments are not automatically capable of articulating societal or policy problems. For example: What would happen if no sediment management actions were taken in a specific river? What kind of problems would then occur? Not only in terms of the more technical problems, such as the accumulation of sediment, but also with respect to the functions that might be affected, such as recreation, safety, transportation by ship, etc.

The following, hypothetical, example was discussed in a SedNet workshop on Societal Cost-Benefit Analysis that was held in Warsaw, Poland on 18 and 19

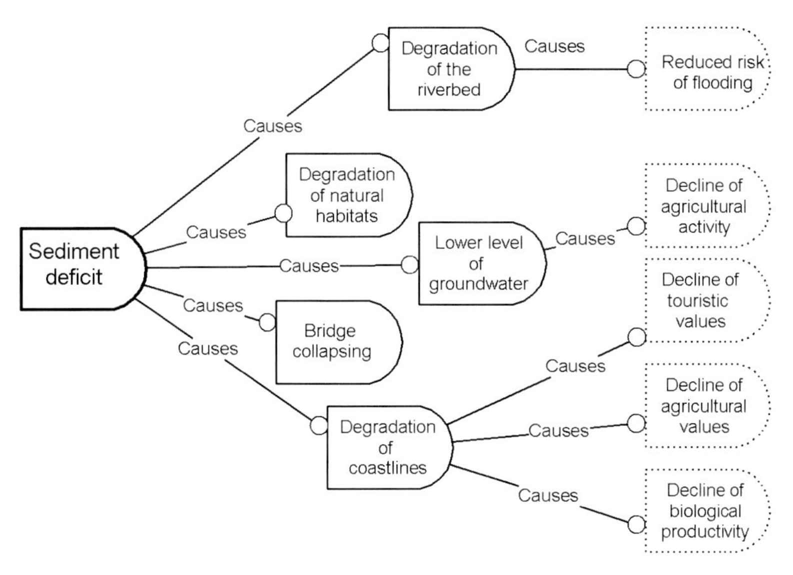

Figure 1. Sediment deficit and its adverse effects

March 2004. It illustrates the difficulties in posing the right questions. The hypothetical problem is a collapsing bridge caused by the erosion of sediment. The sediment management plan should undergo a SCBA and is aimed at minimizing the risk of the bridge collapsing. The description of the problem is the first important step in a SCBA and, therefore, a causal scheme was composed of the possible negative effects of sediment deficit in rivers (see Figure 1).

It is shown that the collapsing bridge is not the only problem related to sediment deficit, but that sediment deficit is related to a whole array of different problems, like decline of biological productivity, decline of agricultural values, etc. So the problem does not become any easier to handle. From the question 'What kinds of impacts does a sediment deficit actually have?', we arrive at the question 'How do we know that there is a sediment deficit?' To answer this question, we should compare the current situation with the 'natural' situation, but this raises the problem of establishing what the 'natural' situation is. It becomes clear that the problem has a temporal scale that has to be taken into account. This raises even more questions, such as 'Is it because of a dam that we have a sediment deficit and shouldn't we therefore evaluate the dam?' This shows that a clear problem description might take quite some time. Knowing the affected values and articulating the policy problems from a broad societal perspective is an essential part of 'understanding the system'. To get this broad societal perspective, stakeholders from different backgrounds and with different interests should be involved (see Chapter 7, this book).

In order to conduct a SCBA it is imperative to have or obtain a very good understanding of the system and its problems. This means that, first, an analysis and an inventory should be made of the problems in the present, and possibly future, situation (see also Chapters 4 and 5, this book).

Second, an analysis should be carried out of the relevant exogenous developments to gain insight into the 'system': the definition of the problem and the evolution of this problem in the course of time if no actions are to be taken.

Third, SCBA is not a content-free, economist tool. Economists often do not have the knowledge that is needed to give content to the method. Therefore, it is necessary to involve all technical expertise, knowledge about eco-systems, and other knowledge from stakeholders. The involvement of stakeholders will not only lead to better knowledge, but is also needed to gain support for the actions that will result out of the SCBA. In other words, a multi-disciplinary approach in an interactive setting with the stakeholders is needed to obtain a clear insight into the existing problems and to conduct a transparent decision-making process.

Fourth, although SCBA includes the application of valuation methods, e.g. to value effects of certain functions like recreation or nature, the tool as such is *not* a valuation method, but an *e*valuation method, weighing alternative actions against each other. This is possibly one of the biggest misconceptions about SCBA. The method needs different actions to tackle a problem. This implies that alternative actions should be generated and the resulting changes should be measured against the so-called 'zero-alternative': 'doing nothing' in an autonomously developing situation. These measurements demand the insights of experts and stakeholders about the system that is influenced.

Fifth, often when the effects of the actions (the changes in well-being and/or welfare) are valued, economists and/or policy-makers are accused of manipulation. It is, therefore, always necessary to specify the effects in their own entities first, before applying any monetary values. The assumptions and methods that are applied to value the actions should be open for discussion. A whole range of methods and guidelines have been developed for this purpose and can be applied in the course of time. When uncertainty is high it may be necessary to apply different methods next to each other and to carry out a sensitivity analysis.

To summarize, the following steps should be taken in a SCBA:

- Analyse the system and define the problem.
- Make an analysis of the evolution of the problem if no actions are taken (the so-called zero-alternative).

- Specify the alternative actions that may be taken to solve the problem (one may use selection criteria such as: the alternatives should not be in conflict with (inter)national laws, should be formulated together with stakeholders, and should be realized within five years, etc.).
- Analyse and specify the possible effects of these actions in terms of changes (compared to the zero-alternative). The effects include possible substitution effects (e.g. the problem at hand may be relocated and can occur elsewhere).
- Indicate the level of uncertainty when predicting these effects (needed for sensitivity analysis).
- Quantify the effects of the actions and value them (not weighing).
- Welfare valuation should be made of all effects irrespective of their nature and irrespective of the place where they occur: always of physical entities (less sediments, more water for irrigation, more nature, etc.) and wherever possible (also) in monetary values. A whole range of different methods exist (Hedonic pricing, shadow prices, travel costs methods, contingent valuation).
- Some effects may occur during a period of time. Discounting rules should be applied to calculate the net present value. (It may be useful to use more than one discounting rate.)
- Compile a SCBA balance sheet with costs on one side and benefits on the other.
- Select and decide (with the relevant stakeholders) what the most appropriate actions are to be taken.
- Decide on extra research questions that have to be answered, develop a monitoring system and conduct any reevaluation.
- In all steps: involve stakeholders (see Chapter 7, this book).

Until now the discussion on SCBA has been mainly theoretical. The following two sections will describe examples of the application of economic instruments in practice. The next section will present a SCBA analysis with respect to sediments in the Netherlands.

4. Example 1: CBA of dredging in the Netherlands

Napoleon Bonaparte called the Netherlands ‘Sludge from the Rhine’. Although intended as an insult, this is an apt description of the Dutch landscape, given the enormous deposits of sediment in the ‘settling basin’ that the Netherlands happens to be. The Dutch waterways support several principal functions, such as recreation, shipping and ecology. To maintain these important functions, dredging is necessary. Although the quality of the sediment is now better, in the 1960s through to the 1980s the sediment was contaminated. This introduced a new problem: increasing costs of dredging and a shortage of disposal facilities for contaminated dredged material. Due to the lack of sufficient disposal capacity (partly as a result of NIMBY: Not In My Back Yard) and increased costs, together with lagging funding, a large backlog was created. These sediments originate both from remedial (environmental) as well as maintenance projects. This backlog was quantified in 2001 and led to the question of whether the benefits of increased dredging, necessary to diminish the backlog, counterbalanced the costs. In Tables 3 and 4 an overview is presented of the major water functions and the dredging volumes.

As shown in Table 3, deferred maintenance of sediment increases by 3.5×10^6 m^3 each year. This will lead to a doubling of the total backlog in about 20 years. The two major functions threatened by deferred maintenance are shipping and agriculture. Contrary to maintenance, the 50×10^6 m^3 for remediation (Table 4) diminishes every year at a rate of 1.3×10^6 m^3. Remediation is closely related to the ecological function of the Dutch waterways, but also relates in part to dredging activities in urban areas (e.g. dredging of canals).

Table 3. Actual vs required maintenance of sediment dredging activities (values $\times10^6$ m^3 $year^{-1}$)

	Required annual dredging	Actual annual dredging	Yearly built up deferred maintenance
Shipping	3.0	1.6	1.4
Ecology	0.1	0.1	0.01
Urban	1.3	0.6	0.7
Agriculture	4.9	3.6	1.4
Recreation	0.03	–	0.03
Total	*9.3*	*5.8*	*3.5*

Table 4. Total deferred maintenance and remediation of sediment (values $\times10^6$ m^3 $year^{-1}$)

	Total deferred maintenance	Remediation	*Total*
Shipping	21	–	*21*
Ecology	6	44	*50*
Urban	6	5	*11*
Agriculture	14	–	*14*
Recreation	39	0.5	*40*
Total	*86*	*50*	*136*

In 2003 the Dutch Ministry of Transport, Public Works and Water Works launched a study of the costs and benefits of increased dredging. The goal of this study is to answer the following two questions:

1. What are the economic costs and benefits of increased maintenance, if annual sedimentation equals annual dredging (reaching equilibrium)?
2. What are economic costs and benefits of eliminating the deferred maintenance and of removing the contaminated sediment in 10, 25 or 40 years (getting rid of the deficit)?

These two questions are graphically represented in Figure 2.

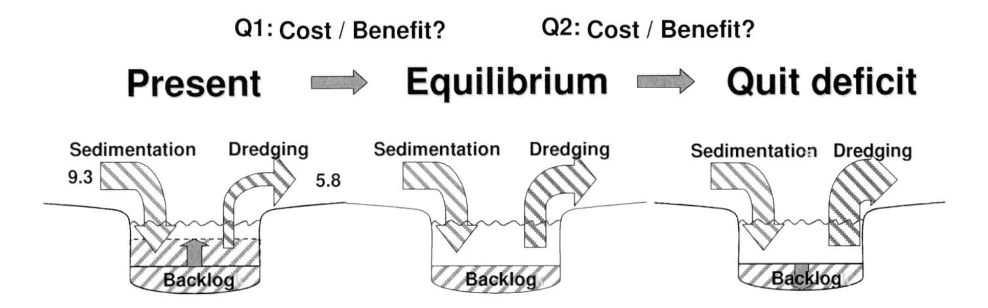

Figure 2. Representation of the two questions in the Societal Cost-Benefit Analysis of the sediment balance of the Dutch waterways

The CBA was carried out following a participatory approach, in which all relevant stakeholders were actively involved, and focused in particular on the following sectors and aspects:

- Shipping: insufficient dredging of waterways results in reduced draught for vessels, implying a reduced load per ship and hence increased transportation costs. This makes shipping relatively less attractive compared to transportation over land, leading to a reduced demand for shipping (modal shift).
- Agriculture: insufficient dredging of regional waters increases the probability of flooding agricultural lands, causing an increase in loss of crop production and eventually terminating the use of land for agricultural purposes.
- Flood hazards: insufficient dredging of rivers increases the probability of flooding the lower elevation polders, unless other measures, such as increasing the height of dykes, are taken.
- Ecology: ecological benefits accrue due to the remediation of contaminated dredging material.

Table 5 shows the costs and benefits of reaching equilibrium and of removing backlogs and remediation in 25 years. In both cases, the net present value is positive. Interestingly, the principal beneficiaries of reaching equilibrium are shipping companies and farmers, while removing backlogs and remediation mainly results in shipping benefits and ecological benefits. This can be explained with the values from Tables 3 and 4, which show that major deficiencies in annual maintenance are in waters with a shipping function or drainage from agricultural areas, whereas the majority of contaminated sediment is found in water with a nature function.

Table 5. Results of the Societal Cost-Benefit Analysis of dredging in the Netherlands (present value in euros×10^9)

Aspect	Reaching equilibrium	Removing backlogs and remediation in 25 years
Costs	1.1	0.4
Monetary benefits		
Shipping	0.9	0.8
Agriculture	0.5	0.1
Reduced flood hazards	0.1	0.0
Total	1.5	0.9
Net present value	0.4	0.5
Other benefits		
Ecological benefits*	5%	20%
Recreational benefits	Positive, but not quantified	Positive, but not quantified
Urban benefits	Positive, but not quantified	Positive, but not quantified

* Compared to present situation.

This study has produced some powerful insights that can enable high and low level decision-making about increasing Dutch dredging activities. It has shown that increasing dredging efforts to reach equilibrium (annual sedimentation equals annual dredging) is beneficial for the Netherlands. The same applies to the economic costs and benefits of eliminating the deferred maintenance and of removing the contaminated sediment in 10, 25 or 40 years.

4.1. Public perception and valuation of biodiversity

In Section 3 we mentioned that a whole range of methods are available for valuation of actions to express the actions in terms of money. One of these methods is the 'willingness to pay' method. This method has been applied in the context of the SCBA presented above. A large scale survey has been carried out in order to assess public perception and valuation of contaminated sediment clean-up and the corresponding positive effects on biodiversity in and around aquatic ecosystems (rivers and lakes) in the Netherlands. The positive effects on biodiversity of this clean-up were assessed by expert judgement. Based on this expert judgement, two possible scenarios of environmental change and the corresponding effects on biodiversity have been developed: a baseline scenario without any additional clean-up efforts; and a policy scenario with additional clean-up efforts.

These scenarios were included in a survey sent to a cross-section of 5,500 Dutch households. The households were asked a range of questions regarding their knowledge and perception of water quality problems in general and contaminated sediments in water systems more specifically. Their attitudes and preferences towards the presented scenarios were also requested. About 1,000 households responded to the survey, providing a rich blend of views and opinions, thereby adding an important public dimension to the overall impact assessment.

Besides providing important indicators about public perception of water quality problems and the need to do something about them, the survey also aimed to estimate public willingness to pay for an increase in biodiversity as a result of increased contaminated sediment clean-up efforts. In this assessment, willingness to pay was used as an indicator of the public non-market benefits of increased clean-up efforts compared to the baseline scenario. In the willingness to pay approach, a monetary value is included for the non-market benefits related to biodiversity preservation and enhancement. This can then be used to see to what extent the necessary investment costs to clean-up the stock of contaminated sediments in the Dutch water system can be justified.

Almost 95% of the Dutch population who responded to the survey indicated that they believe it is important to increase clean-up efforts for contaminated sediments in aquatic ecosystems, and 75% of respondents are willing to pay extra for this as well. Average willingness to pay ranges between 10 and 50 euros per household per year. Relating this amount to their actual annual water bill, this corresponds to a maximum increase of 10% over the next 10 years. Using a conservative aggregation and estimation procedure, and expressed in terms of present value for the cost-benefit evaluation, the total economic value equals 523 million euros.

5. Example 2: Economic analysis and river basin management in relation to the EU Water Framework Directive

In this section an example is given of the application of economic instruments to water management, for which the Water Framework Directive (WFD) is an important driver. It provides a good example of what could be applied to sediment management. The WFD [11] 'aims to establish a framework for the protection of inland surface waters, transitional waters, coastal waters and ground waters'. Its specific purposes are defined in Article 1, as:

1. to prevent deterioration of, and where necessary enhance, the status of aquatic and related ecosystems;
2. to promote sustainable water use;
3. to aim to progressively reduce, and for priority substances eliminate, pollution from hazardous substances;
4. to ensure reduction/prevention of groundwater pollution; and
5. to contribute to the mitigation of floods and droughts (see also Chapter 3, this book).

Although ecological quality is the main criteria to judge the quality of water ecosystems, the main purpose of the WFD is to contribute to the 'provision of a sufficient supply of good quality and quantity of water services as needed for sustainable, balanced and equitable water uses' [11]. The WFD recognizes that water uses by the economic system determine the ecological quality of water ecosystems and, therefore, that influencing water uses is a key to sustainability. Decisions on water management must take into account benefits and costs to society. In this way, water ecosystems are viewed as part of society's natural capital that must be managed in a sustainable way: preserving the integrity of the water environment and its associated ecological functions is the only way to guarantee the services provided by them for the economic system. Water ecosystems provide scarce water services to many conflicting societal and economic targets, and the Directive promotes and sets a common framework for economic principles to guide decisions in four main respects:

1. the valuation of water services and its alternative uses (Article 5);
2. the identification of costs of the provision of water services having regard of the polluter pays principle and the efficient use of them (Article 9);
3. the use of economic instruments to achieve the desired objectives, including pricing and market mechanisms (Article 11); and finally
4. the use of economic appraisal methods to guide the water resource management decision-making process.

In this section we mainly address the last of the aspects mentioned above and more specifically the role of CBA in general, and cost effectiveness analysis in particular, as decision support tools in the implementation process of the WFD. Although the use of economic principles for water management has been encouraged for a long time, and its importance has been enhanced with the appearance of water demand instruments, the WFD approach is highly innovative in many respects. The economic analysis of the WFD implies an important challenge because of the lack of proper information systems and databases and also due to the lack of tested methods for empirical applications. River basins are complex ecological systems with many interactions that need to be taken into account (see Chapter 1, this book). Sediment balances depend on the river hydromorphology (see also Chapter 4, this book). Water quality and quantity are closely linked to each other. The connections between runoff and underground water are not well known. There are many competing water uses for any water body, and present decisions on water abstraction have uncertain effects on future welfare. Additionally, although important progress on valuation methods of water services has been attained in previous decades, results are not completely robust and there is still some important discussion on the proper way to integrate the value of the many water services implied in a common integrated framework. Two points must be made to understand how the WFD copes with these problems in a practical way.

First of all, (socio-)economic analysis and SCBA must not be interpreted as a decision-taking framework but as useful tools in the decision-making process. For that reason the economic analysis needs to be integrated with other expertise and analyses in supporting the development of river basin management plans. Efforts into more detailed economic analysis should be proportionate and concentrate only on significant water management issues, areas of conflicts between uses and where the integration between environment, economic and societal issues is problematic; in other words, where economic analysis can help in making better decisions.

Second, economic analysis should serve to improve the quality of the societal decision-making process, informing about possible policy choices or helping to justify these choices and conveying information to the public/stakeholders. Involvement of stakeholders into the economic analysis is a way to bring expertise and information, provide opportunities to discuss and validate key assumptions, and to increase societal involvement and the acceptance of the results of the economic analysis (see Chapter 7, this book). According to the WFD, economic analysis should report on information, assumptions and approaches used for obtaining results in a transparent way as a prerequisite to enhance information and participation of the stakeholders.

To understand the function of CBA in the water management decision process, it is important to compare the abstract theory of SCBA with the way this theory is transformed into practical recommendations to implement the WFD. Ideally, SCBA is a method to solve different policy problems in a common framework. A necessary condition for applying SCBA is to solve two basic informational problems. The first one refers to the benefits associated with an improvement in the ecological quality of a water body, and the second one refers to the question of the opportunity costs to obtain such an improvement.

The benefits are represented as the welfare gains that result from both the increased capacity of the water ecosystem to provide services to the economic system (for instance, improved recreational services, reduced flood risk or a higher guarantee of water supply in dry periods), and the preservation of existence and option values (associated with a higher biodiversity, increased options for future generations, etc.). Figure 3 illustrates the marginal benefits of improving water quality as a decreasing curve, showing that the quality is improved at a decreasing rate. To capture these benefits, society must 'invest in' or pay the opportunity cost of improving the ecological quality. These costs come from the investment and operation expenses of the measures needed to improve water quality and also from the negative effects of reducing the economic pressures on the water ecosystem (or by reducing water abstraction, pollution, economic uses of head waters, etc.). The hypothetical marginal cost of ecological quality is represented in Figure 3 by the increasing curve.

If we were able to gather all the information about costs and benefits associated with an improvement in the ecological quality, the application of CBA would become a relatively easy task. In that case, society must simply try to improve the water quality up to the point where the cost of improving quality in the margin is not higher than the benefits society can obtain from this change. Ideally, CBA would allow us to solve many economic questions at the same time. First, as we have seen, the method provides a way to define the optimal ecological quality. Second, it is also a way to measure the environmental gap that needs to be closed to reach that optimal target. This gap is simply the difference between the current and the optimal ecological status. Third, it provides a method to measure the welfare gains that may be obtained when the optimal status is reached (represented by the triangle WG in Figure 3). And finally, it is a way to identify the actions that should be taken to obtain the optimal status at the least cost to society (all the policy measures having a marginal cost lower than the marginal benefit of the ecological improvement measured at the optimal ecological quality) and to design the optimal river basin management plan.

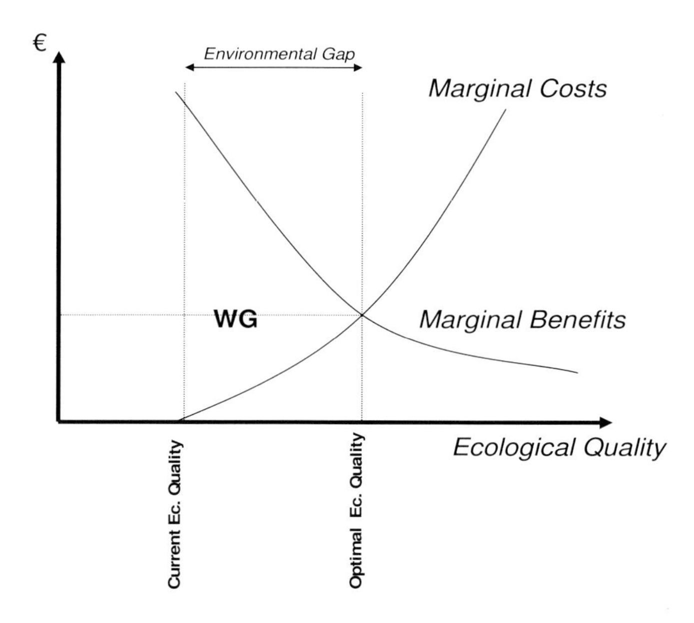

Figure 3. The theory of Cost-Benefit Analysis applied to the EU Water Framework Directive (see text for explanation)

Nevertheless, recommendations from abstract theory are rarely easy to put into practice. Experience in water management has led to the conclusion that the costs of the actions needed to improve ecological quality are easy to value in a simple and precise way. Contrary to costs, the benefits of ecological improvements are diverse, including the welfare gains to landscape, recreation, safety, biodiversity and so forth. Most of the benefits are local and their values can not be easily transferred from one ecosystem to the other. Many of them are qualitative and difficult to value in welfare terms, different cost valuation methods lead to different values and the values obtained are always uncertain.

To cope with these informational restrictions, and to make economic analysis feasible in the practical arena, the WFD assumes that societal decisions on water management must be taken in an iterative institutional process, where the starting point must be the setting of a desired ecological status on the basis of technical knowledge and stakeholder involvement. These preliminary objectives may be refined in a more advanced phase of the WFD implementation when SCBA will play a key role.

The decisions sequence may be represented by a process including the following main phases. First, a preliminary target of a good ecological status of the different water bodies or a water basin is defined. Second, the combination of policy measures to obtain the desired target with the least cost to society is determined. Third, the benefits associated with the ecological improvement are identified and described, but not valued. Fourth, the measure of costs and the benefits are presented to stakeholders in order to determine whether benefits are perceived as high enough to justify the effort needed to obtain the good ecological status. If the answer to the previous question is affirmative, the analysis proceeds by defining the institutional constraints, the distributive effects of the river basin measure package, the financial constraints and the many other aspects that need to be dealt with to implement the river basin plan. If costs are perceived as disproportionate, in the WFD jargon, the desired ecological status is modified by setting a less stringent ecological status (a target derogation), or by allowing more time to reach the good ecological quality (a time derogation).

One specific tool to analyse a way of reaching the predetermined 'good ecological status' is Cost Effectiveness Analysis (CEA). In the WFD the identification of the welfare gains of improving the ecological quality of a river basin is not used to set the water management policy objectives, but to assess whether the benefits are higher than the costs associated with the best alternatives to obtain the predetermined good ecological status.

Once the WFD is fully implemented, CEA will play an important role in the definition of an adequate combination of policy options that may be chosen to achieve a desired ecological status in any European river basin. Nevertheless, CEA is just an intermediate stage in the design of a river basin management plan, and should not substitute the decision-making process itself. It takes information produced by previous economic analysis, based upon calculation and estimation of costs and physical effectiveness of identified measures to meet a given standard as, for example, the Good Ecological Quality (GEQ) or the Moderate Ecological Quality (MEQ) of the river basin. The output(s) of CEA are especially relevant since an ultimate CBA (as well as the consultation process) that would lead to final decisions about objectives, timing and required measures, will need to be developed on that basis. The CEAs prepare the ground for the SCBA with a wider scope.

The rationale of the CEA is shown in Figure 4. This figure shows the marginal cost (MgC) of achieving a given level of a parameter of environmental quality (Q). We can think of Q as measuring a given attribute, such as the concentration of a specific pollutant, water temperature, the rate of flow, etc. The marginal cost curve reflects the supply side of providing a better environmental quality and can be derived through ranking all the alternative policy options according to the marginal cost of providing an increase in parameter Q. This type of analysis can also be applied to sediment management, whereas sediment balance is only one parameter for ecological quality. Figure 5 illustrates a hypothetical example where the environmental quality is a measure of the distance of a sediment balance with respect to a natural regime.

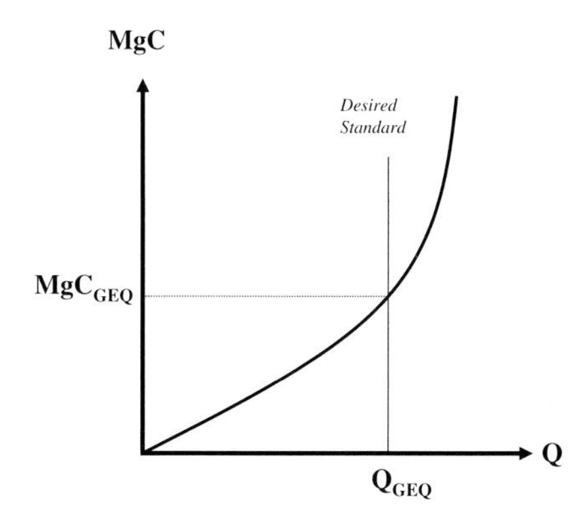

Figure 4. Cost effectiveness analysis (see text for explanation)

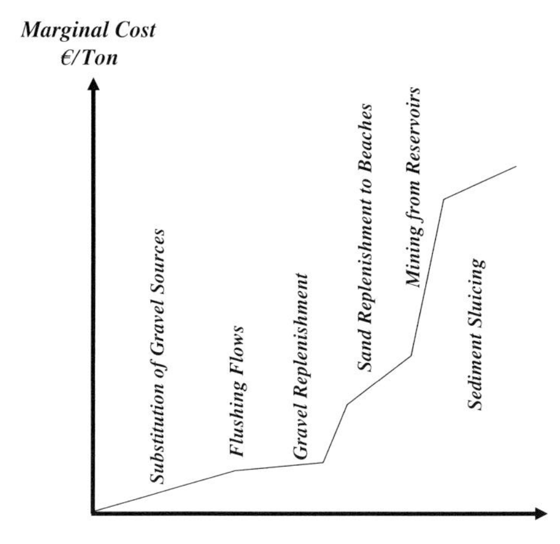

Figure 5. Example of marginal costs of restoring sediment balances

There are several options to improve the sediment balance. As an illustration, the following list contains the most representative examples of restoration and remedial measures to improve sediment balances, based on the literature:

1. Substitution of gravel mining sources. Designed to reduce or eliminate instream gravel mining by defining a minimum elevation of the thalweg or by allowing extraction of only a fraction of the natural annual bedload

sediment supply. This will increase the cost of sediment as an input for construction and other economic activities.
2. Flushing flows. A measure designed to partially remove fine sediments accumulated on the bed and to scour the bed frequently enough to prevent encroachment of riparian vegetation and narrowing of the active channel [12].
3. Gravel replenishment below dams. Artificially added gravels to enhance available spawning gravel supply can provide short-term habitats. Experience shows that imported gravels are highly mobile and thus measures need to be taken on a regular basis, depending on the magnitude of the runoff [13].
4. Sand replenishment to beaches from headwaters, river transport and offshore sources. Designed to compensate sand supply reductions to beaches due to a reduction in sediment delivery from streams and to avoid increased risk of shrinking and cliff erosion.
5. Mining aggregate and industrial clays from reservoirs. Although financially expensive, this measure may be economically viable when the benefits of increasing reservoir capacity and the environmental benefits of reduced instream and floodplain mining are taken into account.
6. Sediment sluicing and pass-through from reservoirs. A measure designed to partially restore continuity of sediment transport. Special care must be taken to control the possible negative effects of an abrupt increase in sediment load on water quality and aquatic habitat conditions downstream.

The application of CEA to this set of measures may lead to a marginal cost curve of improving sediment balances, as is represented in Figure 5.

The main challenge for the implementation process of the economic analysis aspects of the WFD is to integrate the many different alternative measures that may be taken in any water body to improve any single quality parameter. The goal is to obtain the least-cost combination of policy measures that guarantee the overall good ecological quality of the entire river basin. In this integration process special care must be given to the following aspects. First, there is the integration of different measures taken at different water bodies to improve a given ecological quality parameter. With respect to sediment management, for example, the river basin can be divided in three zones: the erosion zone (typically the headwaters), the transport zone and the deposition zone (for details see Chapter 4, this book). By planning at the basin level, the cost of, for example, improving continuity between the erosion and the transport zone will reduce the need to reduce sediment extraction in the deposition zone. In other words, part of the marginal restoration cost of restoring continuity is compensated for by the avoided compliance costs downstream.

Second, there is the integration of different measures performed to improve different quality parameters. In this case, improving sediment balances, for example, will have an internal effect in the river basin, by enhancing available spawning areas, thus reducing the cost of reaching the good ecological status. In another example, flushing flows may prevent the encroachment of riparian vegetation avoiding other restoration measures. Measures intended to restore sediment balances may substantially improve water quality. In sum, the integrated cost effectiveness analysis proposed by the WFD provides the tools needed to appreciate the benefits of any single measure in terms of the avoidance of the societal cost of guaranteeing the desired ecological status of the river basin.

6. Environmental liability and sediments

Next to the WFD, described in the previous section, another directive from the EU that could have a great impact on sediment management is the Environmental Liability Directive (ELD) (also see Chapter 3, this book). The ELD was approved by the Council and the European Parliament on 30 and 31 March 2004, respectively. The Directive will enter into force on the day of its publication in the Official Journal. The ELD aims 'to establish a common framework for the prevention and remedying of environmental damage at a reasonable cost to society' [14]. The Directive not only covers damage to persons or goods and contamination of sites, but also damage to nature, especially to those natural resources that are important for the conservation of the biological diversity in the Community. The ELD wants to reach its goal by implementing the 'polluter pays principle', which ensures that whoever causes environmental damage pays to remedy the damage. To ensure an effective implementation of this Directive, persons or non-governmental organizations adversely affected or likely to be adversely affected by environmental damage should be entitled to ask the competent authority to take action. The EU deems that because environmental protection is a diffuse interest, non-governmental organizations promoting environmental protection should be given the opportunity to properly contribute to the effective implementation of this Directive (Directive 2004/35/CE, note 25).

This Directive will have a great impact on sediment management, and the costs and benefits that have to be taken into account, because organizations in addition to those concerned with environmental protection will have a strong instrument to emphasize the importance of sediments in the ecosystem. Government organizations and companies have to be very careful in influencing the quality and quantity of sediments, as they will be accountable for their actions. This means that, in the end, they will have to pay the bill to restore the quality of the ecosystem. For example, when a government organization decides to build a hydroelectric dam, which would have a great impact on sediment balances, the government organization is obliged to make amends to restore this. Another aspect of the ELD, which is of great importance especially to sediments, is the polluter pays principle. Thus far it was sometimes difficult, if not impossible, to track the sources of certain sediment contaminants, especially in rivers that ran across national borders. This meant that the government organizations that had to deal with the contamination also had to pay for the remediation. However, with the implementation of the ELD, the development of techniques that can be used to track the source of the pollution (see Chapters 4 and 5, this book) deserves more attention, as these techniques would enable government organizations to actually claim the costs.

The ELD, therefore, provides a strong basis to raise the awareness of organizations of the consequences of their actions. Environmental liability with respect to damage to nature is a prerequisite to making economic groups feel responsible for the possible negative effects of their operations on the environment.

7. Conclusions

This chapter has given a short introduction into the methodology of SCBA in relation to sediment management. SCBA is an elegant tool to evaluate different policy options based on a rational approach. One has to bear in mind, however, that human decisions are not very rational, and that rationalities between different groups will differ. To obtain meaning from a SCBA to these different rationalities, it is necessary to involve stakeholders in the steps of the SCBA. The first step in a SCBA is to do a system analysis and to define the problem. As we have shown in this chapter, this is not as easy as it may seem. First, different groups will have different views on the problem, but most of all, problems in a natural system will have relations to each other, so solutions will have relations to different problems (in different ways). An assessment of the problem(s) and solutions will help to gain a better insight into these relations. The various steps in a SCBA are presented in Section 3 and the example of sediment dredging in the Netherlands showed these steps. For the valuation of the different actions that can be undertaken, different methods are available. In the Dutch example, the application of the 'willingness to pay method' was shown. Overall, the Dutch SCBA is a nice example of the application of SCBA to sediment management and produced some powerful insights that helped decision-making about Dutch dredging activities. The example of application of economic analysis instruments to water management options shows the influence of the EU WFD on the application of economic instruments and illustrates how cost effective analysis can support decision-making.

Whether we use a SCBA or other economic analysis instruments for decision support, we should be aware of the limitations of these methodologies. Nevertheless, SCBA can be a very useful and powerful tool to help societal decision-making and to facilitate the more efficient allocation of society's natural resources, of which sediment represents one type. The following chapter (Chapter 7, this book) considers in more detail the specific role of stakeholders in the decision-making process for sediment management.

References

1. Turner RK, Pearce D, Bateman I (1994). Environmental Economics: An Elementary Introduction. Harvester Wheatsheaf, New York.
2. Brent RJ (1996). Applied Cost-Benefit Analysis. Edward Elgar, Cheltenham.
3. Boardman AE, Greenberg DH, Vining AR, Weimer DL (1996). Cost Benefit Analysis: Concepts and Practice. Prentice-Hall, London.
4. Bouma JJ, Saeijs HLF (2000). Eco-centric cost-benefit analysis for hydraulic engineering in river basins. In: Smits AJM, Nienhuis PH, Leuve RSEW (Eds), New Approaches to River Management. Backhuys, Leiden, pp. 167–178.
5. Self P (1975). Econocrats and the Policy Process: The Politics and Philosophy of Cost-Benefit Analysis. Macmillan, London.
6. Lohmann L (1998). How Opinion Polling and Cost-Benefit Analysis Synthesize New 'Publics' (source: http://www.thecornerhouse.org.uk/briefing/07cba.html).
7. Kelman S (1981). Cost-benefit analysis: an ethical critique. AEI Journal on Government and Society Regulation, January/February: 33–40.
8. Bouma JJ and Schuijt K (2001). Ecosystem valuation and cost-benefit analysis as tools in integrated water management. In: Midttømme GH, Honningsvåg B, Repp K, Vaskinn K, Westeren T (Eds), Dams in a European Context. Swets & Zeitlinger, Lisse, pp. 154–163.
9. Stuip MAM, Baker CJ, Oosterberg W (2002). The Socio-Economics of Wetlands. Wetlands International and RIZA, The Netherlands (source: http://socioeconomic.unep.net/reports/SocioEconomics_of_Wetlands.pdf).
10. Susskind L, McKearnan S, Thomas-Larmer J (1999). The Consensus Building Handbook: A Comprehensive Guide to Reaching Agreement. Sage, Thousand Oaks, CA.
11. Directive 2000/60/EC of the European Parliament and of the Council establishing a framework for the Community action in the field of water policy.
12. Reiser DW, Ramey MP, Wesche TA (1989). Flushing flows. In: Gore JA, Petts GE (Eds), Alternatives in Regulated River Management. CRC Press, Boca Raton, FL, pp. 91–135.
13. Kondolf GM, Kattelmann R, Embury M, Erman DC (1996). Status of riparian habitat. In: Sierra Nevada Ecosystem Project: Final Report to Congress, vol. II, Assessments and Scientific Basis for Management Options. Wildland Resources Center Report no. 37. University of California, Centers for Water and Wildland Resources Davis, CA, pp. 1009–1030.
14. Directive 2004/35/CE of the European Parliament and of the Council of 21 April 2004 on environmental liability with regard to the prevention and remedying of environmental damage.

Sustainable Management of Sediment Resources: Sediment Management at the River Basin Scale
Edited by Philip N. Owens

Sediment Management and Stakeholder Involvement

A.F.L. Slob[a], G. J. Ellen[a] and L. Gerrits[b]

[a]TNO Environment and Geosciences, P.O. Box 49, 2600 AA Delft, The Netherlands
[b]Erasmus University Rotterdam, Department of Public Administration, P.O. Box 1738, 3000 DR Rotterdam, The Netherlands

1. Introduction

Stakeholder involvement will be of increasing importance for environmental policy-making, not only because European Union regulations demand societal participation, but also because of the increasing complexity of environmental policy issues in general. This complexity is caused by the many groups that have a role in environmental problems, the competing interests of stakeholders and the involvement of several policy levels (regional, national and international). When looking at the sediment issue on a river basin scale, the complexity is obvious (see Chapters 1 and 2, this book). There are many stakeholders as well as policy levels that are involved within river basins (see Chapter 3, this book) and the interests of all these groups are quite diverse. This chapter will focus on the basic questions concerning stakeholder involvement in the decision-making processes with respect to sediment management and, consequently, on the recommendations that can be derived from this. The chapter is based on both the scientific literature and the SedNet workshops.

In contemporary European and American policy-making, an increase of interactive processes and stakeholder involvement in relation to policy-making can be observed [1–11]. These processes bear different names, like interactive governance, co-production, participatory processes and so on. In the relevant academic and professional literature, many definitions and descriptions of stakeholder involvement can be found [1, 2, 12]. The core theme in those definitions is that governments develop policies in consultation and cooperation with stakeholders, as defined above. Edelenbos defines stakeholder involvement as 'the early involvement of individual citizens and other organized stakeholders in public policy-making in order to explore policy problems and develop solutions in an open and fair process of debate that has influence on political decision-making' [13]. Stakeholder involvement as a process differs from traditional public consultation procedures in that stakeholders are involved early enough to influence policies when they are formulated. This is opposed to the classical European approaches to public decision-making where decision-making power remained firmly with the representatives.

The structure of this chapter is as follows. First, attention will be given to the question of *who* the stakeholders are (Section 2) and *why* stakeholders should be involved (Section 3). Stakeholder analysis is a method to make an inventory of all stakeholders that have a role (a stake) in the issues and solutions. This is a necessary step to know which stakeholders one should involve in sediment management and will be discussed in Section 4. Stakeholders should not be considered as one, homogenous, group. Stakeholders are diverse and heterogeneous and this is reflected in the way they look at sediments (Section 5). This inevitably leads to discussion on *how* to involve stakeholders (Section 6) and the tools and mechanisms to do this (Section 7). The last section (Section 8) deals with the risks and pitfalls that are attached to stakeholder involvement, for which the policy-makers should be warned. After all, an interactive process in which stakeholders are involved is not as straightforward as it may seem.

2. Stakeholders

Hahn [14] provides a useful scientific overview of the evolution of the stakeholder concept. He refers to Freeman [15], who popularized the concept and initiated further development in the many subsequent publications (for an overview on the stakeholder literature, see Mitchell *et al.* [16]). In Hahn's view the stakeholder concept takes up an actor perspective and analyses the exchange relationships between companies and various groups that have some kind of stake or interest in a company or issue. He states that a common definition of the term 'stakeholder' is still lacking. Hahn [14] distinguishes between narrow and broad definitions: 'According to the broad definition stakeholders are all those groups and individuals who are affected by and/or affect corporate goals [15]. However, the broad definition has been criticized because it provides little guidance for the distinction between stakeholders and non-stakeholders. There are various narrow approaches to identify and define stakeholders. These can be grouped into normative and instrumental approaches [16]. Normative approaches use moral criteria in order to focus on those that should be considered as stakeholders. By doing so they put the moral justification of the relationship between a company and another actor to the fore [17]. Instrumental approaches define and identify stakeholders by their relevance for corporate success. Thus, stakeholders are all those actors without whose contribution the company could not be operated successfully [18] or would even cease to exist [15].' Hahn then points to the importance of stakeholders as resource providers with respect to this, and on the management of stakeholder relations. He argues (based on [19]) that for successful stakeholder management, stakeholder–company relationships have to be analysed as a network of various actors influencing each other rather than as dyadic ties. He distinguishes between direct and indirect stakeholders according to the way they have an impact on corporate success: direct stakeholders are involved in direct exchange relationships with a company, whereas indirect stakeholders influence such direct stakeholders or exchange relations.

These definitions might seem a little abstract when applied to sediment management, because they are directed towards companies and their stakeholders. But the definitions can easily be translated towards public organizations making decisions on sediment management. In one of the SedNet workshops, dealing with stakeholder involvement, the definitions were tailored to sediment issues. A distinction was made between three types of stakeholders:

- Organizations and people that have a direct impact on sediments or are directly affected by the relevant policies. This group includes: harbour authorities; shipping companies; dredging companies; industries using water and/or dumping their wastewater; farmers; water authorities; water cleaning companies; regulators on the local, regional, national and international level dealing with water issues and subjects of environment, agriculture and safety, or with conventions such as OSPAR and HELCOM; maritime organizations; international river committees; organizations maintaining natural defences; water managers; owners of nature areas; and citizens that are directly affected by the measures planned or taken.
- Organizations and people that have an impact on the relevant decision-making. This group covers: citizens; landowners; homeowners; insurance companies; NGOs such as Greenpeace and the WWF; scientists; and drinking-water companies.
- Those who have an indirect impact on or are indirectly affected by sustainable sediment management: this group consists of all the other users of the waterways and fisheries, and includes citizens.

3. Why stakeholders should be involved

There are several arguments for stakeholder involvement with respect to sediments. Apart from the basic fact that stakeholders have an impact on the quality and quantity of the sediments along the river, the main arguments can be grouped into three themes: obstructive power; enrichment; and fairness [20].

In modern society, parties other than governments have obstructive power. That is, they have the ability to obstruct or even block a decision or the implementation of a certain policy. Deposit sites for dredged material, for example, always alarm citizens living near the site. They can, and often do, protest against it or take other measures. The early involvement of stakeholders reduces the risk of the policy not being carried out. Stakeholder involvement, therefore, can be regarded as counteracting obstructive power [1, 2]. Such a choice will slow down the policy process in the early phases, but will speed it up in a later phase.

The abovementioned reason for stakeholder involvement is sometimes regarded as a negative one, born out of strategic considerations. However, there is a positive motive as well. This considers enrichment. Governments do not possess all the resources, i.e. knowledge, required for the design, planning and implementation of sophisticated policies such as environmental policies. Relevant knowledge is in most cases distributed among several stakeholders. This counts the more for sustainable management of sediments, a subject where knowledge is still fragmented and debated. From that point of view it is wise to invite stakeholders from the relevant fields in order to obtain and apply knowledge and information generated by them [8]. No one can provide as much local insight to aid planning for the development of, for example, a dumping facility for dredged material as the local dredging companies, the people living in the vicinity of the site and the pressure groups that work to protect the natural and human environment in the area [20]. Stakeholder involvement can provide good ecological practices in this way.

The last argument for stakeholder involvement is fairness. It is fair to involve actors affected by a certain policy, such as the construction of a dumping site, and give them a say in the decision-making process. Politicians are often in favour of this argument, especially when it comes to controversial topics such as the dumping of contaminated dredged material. Norris [21] shows a global increase of support for such democratic arrangements. In addition, Van Ast [22] shows that internalization of sustainable behaviour among stakeholders of the river can only be reached through the involvement of those actors in the policy process. This raises awareness and creates support for the issue and its solutions.

As a final remark on the motives for stakeholder involvement, we want to point out that stakeholders, especially the organized ones, often will look for ways to get involved themselves, as they are aware of their stake in the process.

4. Stakeholder analysis

The first step in the process of involving stakeholders is often referred to as convening, i.e. getting people to the table. This consists of four steps: assessing a situation (convening assessment); identifying and inviting the stakeholders;

locating the necessary resources; and organizing and planning of the process [23]. We will focus here on the identification and analysis of the stakeholders.

Stakeholder and network analysis starts with assembling a list of all relevant stakeholders. This can be done with a small group of people belonging to the convening organization, who know the issues and have an overview of possible stakeholders. This group should be diverse, as this prevents a one-sided selection of stakeholders. The first step in the process is to identify the different perspectives on the issue with a wide variety of people. At this stage of the process, the goal is not to identify people or organizations who should be involved. At this point, the way the different stakeholders look at the issues and solutions is categorized. The second step is to see which sub-perspectives can be identified. The third step is to identify different key persons or organizations within the identified sub-perspectives. The result of the last step will be a list of stakeholders. An example of this first phase of the stakeholder analysis is given by the National Research Council [24] who, dealing with a community around a contaminated site, roughly identified the following stakeholders: 'The community surrounding a contaminated site includes the people who live closest to the site and whose health may be at risk and/or whose property, property values or economic welfare is adversely affected by the contamination; local business owners; elected officials; local government agency representatives; workers at the site; and others who live farther from the site but who are indirectly affected' [24].

The list of stakeholders is the steppingstone for the next phase of the stakeholder analysis: to identify the interests and goals of these stakeholders in the process. For each stakeholder the following questions should be addressed [25]:

- What will the stakeholders contribute to the process?
- What kind of knowledge do they possess?
- What are the relevant interests and goals of the stakeholders?
- How do the stakeholders interpret the issue at hand?
- How well-informed are the stakeholders about the issue?
- What are the (possible) motives for these stakeholders to participate, or not to participate?

The majority of these questions can be answered based on experience or policy documents. However, it will always be necessary to do a number of interviews with stakeholders to get more background information. The benefit of this is that the interviews will create a means for people to voice their concerns about the issue, which they might be reluctant to do when confronted with other stakeholders [23].

The final phase in the stakeholder analysis is to make an inventory of the relations between the different stakeholders. This is necessary to understand certain attitudes or actions of stakeholders concerning sediment. Relations can be identified by desk research, but also by asking the stakeholders themselves. It is usually very helpful to visualize relations between stakeholders and to identify them as positive or negative for the process. By making this transparent, it is possible to deal with negative relations and to use positive relations to improve the quality of the process.

5. Stakeholder perspectives

Stakeholders do not necessarily share the same view or perspective [26, 27]. With perspective we mean 'a consistent and coherent description of the perceptual screen through which (groups of) people interpret or make sense of the world and its social dimensions, and which guides them in acting' [28]. This phenomenon can be explained by the fact that every individual has their own set of values that works as 'a pair of glasses' through which the world is perceived, which means that situations and occurrences are interpreted differently. The individual's argumentation and rationality are influenced by an individual's perspective. The consequence is that people select different phenomena for their assessment and, when choosing the same phenomena, they most likely interpret these phenomena differently by putting different values on them [29].

Because of the fact that the perspectives of stakeholders plays such a guiding role in the actions they undertake, the perspectives of actors with respect to sediment management were investigated. In a case-study concerning sediment issues, we conducted a number of in-depth qualitative semi-structured interviews with stakeholders. In these interviews, we asked the stakeholders about a number of important views on sediments, for example what the 'perfect' river basin would look like according to them and how they prefer to solve problems related to sediments. Based on the analysis of these interviews, we came to the conclusion that we could distinguish three different perspectives on sediments. To validate the outcomes, the stakeholders were given the opportunity to give feedback on the perspectives. Together with our analysis, this resulted in the description of the perspectives that are shown below [30].

1. *Users*. This perspective is characterized by a short-term vision of sediment. From the Users' perspective, sediment can be seen as a useful resource, for example, as a fertilizer, building or construction material or a resource for elevating land. But sediment can also be an obstacle, which needs to be removed to make waterways more accessible for recreational and/or commercial shipping. The sustainable management of sediment in this

perspective is focused on creating (economical) possibilities. To do this in an optimal way, technology is often applied. The Users' approach to policy and regulation is of a practical nature; every situation needs to be viewed in its own context, and regulation should be applied accordingly.
2. *Controllers.* The most important aspect of sediment from the Controllers' perspective is to avoid societal risks. Sediments can pose a threat to society, mainly in two ways: flooding and pollution. From the Controllers' perspective it is, therefore, very important to control both the risks of flooding and the risks of pollution for environmental and public health. The timeframe is set for the short-to-medium term, so the controllers anticipate future developments, but also keep a close eye on the current situation (focus on risk management). Information gathering and research are of great importance for this perspective. Research is used to deal with uncertainty of any kind. There is never enough information on an issue. Policy and regulation are seen as a framework for 'keeping control' over a seemingly controllable and rationalized situation.
3. *Guardians.* Guardians operate with a long-term timeframe in mind. Guardians see sediments as part of the ecosystem that should be handled with care. Changing the present situation can have consequences, sometimes unintended, for the development of the ecosystem. The focus is mainly on the quality of sediments. Safety of the society is also important, but only when no alternative is available is interfering with the ecosystem allowed. If there is a situation in which this is unavoidable, then it should be done with the utmost caution. The Guardians' goal is to obtain 'natural' sediment, because this is an important aspect of the ecosystem, which influences water quality and the variety of flora and fauna. Policy and regulation should be directed towards making this goal attainable and towards preventing short-term (economic) goals from intervening.

Insight into the actual perspectives is necessary when communication between, or within, groups takes place. As their perceptions and rationality differ, the concepts and words used, i.e. their language, will also differ. So they speak different languages. This means that exchange of concepts and words will not automatically lead to understanding and mutual comprehension of sediment problems.

In Table 1, an analysis is presented of the language that is used by the different perspectives. The words and concepts that mean a lot to the various perspectives are identified. The 'counterpart' of that, the words that have no or minor meaning in this particular perspective, are called the 'blind spots'. As these words do not have any meaning in a perspective, they should preferably not be used.

Table 1. Language and blind spots of the different perspectives on sediment

Perspective	Language	Blind spots
Controller	Danger/safety of sediment Flooding caused by sediment Sediments influence the quality of the environment and nature Regulation Research on every aspect of sediment Government control of sediment regulation	Long-term impact Experimental or 'risky' solutions will be overlooked Ownership of solutions with stakeholders (the controller wants to keep control over the solutions)
Guardian	Ecosystem: harm to flora and fauna Waste sediments are contaminated and are damaging the ecosystem Risk oriented: polluted sediments can harm the ecosystem Regulation to prevent damage	Economically viable Efficient solutions Costs are not important
User	Challenge and profit Sediment problems can be solved using technology and treatment Pragmatic: if sediments cause problems they have to be managed Costs: the management of sediment can be costly	Long-term impact Ecosystem is not an issue because technology can solve all problems, and the ecosystem will restore itself Risk issues can easily be solved Control and regulation is not necessary

One important observation from Table 1 is that sometimes one perspective's language is the blind spot of the other. For instance, Guardians have a blind spot for economic reasoning (economically viable, efficient), while this is the language of the User. Users do not like regulation and control, while the other two perspectives do. This explains why it is sometimes so hard to get good communication – and mutual understanding – between groups with different perspectives. This stresses the importance of taking time to develop a context in which groups can get a better understanding of each other's interests and to develop a new, common, language.

6. How to involve stakeholders?

The process of stakeholder involvement should require an independent chairperson or process manager. Such a person should not be attached to the involved parties and should be as independent as possible. The arrangement for their payment, for instance, should reflect that. They should be paid by a mixture of stakeholders in order to avoid the appearance of a conflict of interests.

The first step for the process manager in the organization of stakeholder involvement is to find out which stakeholders should be involved. For the stakeholder selection the following questions should be answered: Will the stakeholders be affected by the policy? Are they the target group of the policy? Do they have the power to obstruct or the means to enrich? The questions that should be posed differ with the aim of the process [13]. A crucial requirement here is variety. Although, at first glance, it seems to make more sense to focus on representativeness, variety should be the guiding thread. This triggers the enrichment of the process and serves as a safety valve against overlooking stakeholders. A good way of assembling a stakeholders panel is to ask other stakeholders who they regard as vital for the process. Through this so-called snowball method, other stakeholders who might have been overlooked initially can be invited to the process.

Next, it is important to collect information about the goals, ambitions and problem definitions (from the various perspectives) of the stakeholders. The process manager should ensure that all these interests are heard and acknowledged in the course of the process. If not, stakeholders might pull out, which could damage the process (see also Section 7). In order to guarantee that all interests are present in the process, the manager should be sure to know them. One must be aware of the difficulties of acquiring the desired information, whereas stakeholders can show strategic behaviour. The stakeholders might want to shield their real interests, as they prefer to hide their agenda. Their real interests can sometimes be obtained better through asking other parties.

The mobilization of the stakeholders is an important issue. Too often, decision-makers feel that the majority of the potential stakeholders lack interest, whereas some with strong but specific interests dominate the agenda. So it is the duty of the manager to let stakeholders realize what is in it for them. Why should they join the process? A sound and deliberate consideration of interests might persuade less-interested parties to join and will be a signal to dominant forces not to overact. Furthermore, awareness and urgency should be created. This can be done by pointing out the drivers behind sediment-related issues. These include the regulations issued by the European Commission such as the ‘polluter pays principle’ in the Environmental Liability Directive (see Chapters

3 and 6, this book). Finally, the fairly technical nature of sediment-related problems such as contamination and morphological change needs translation. Laymen and the public cannot be expected to fully understand the technical backgrounds of the problem and, therefore, communication must be clear and free of jargon.

The processes of involvement can be arranged at different levels [20]:

- Information: providing information to the stakeholders.
- Consultation: consulting what stakeholders think must be done.
- Advising: letting stakeholders advise on the policy and taking their recommendations into account.
- Co-producing: stakeholders are regarded as equal policy-makers but decision-making remains in the political domain.
- Co-deciding: decision-making power is handed over to stakeholders.

Every situation is unique and, therefore, the level should be chosen that fits the specific situation. It is most importanct that, once a level is chosen, this is communicated to stakeholders and that the level is not abandoned in the course of the process. Doing so will create uncertainty and distrust. It should also be understood that not every stakeholder has to be involved at the same level. Some just want to stay informed, whereas others want to be heavily involved.

A common feature of processes that bear some kind of uncertainty is that they are often made resistant against unforeseen events. This should not lead to a process of stakeholder involvement where stakeholders cannot enter the process in later stages. A certain amount of openness is required, and openness has two dimensions. The first is openness with respect to new stakeholders. Once the process is on its way, new stakeholders should still be able to participate. At the same time, this should not mean that the process restarts over and over again. The second dimension of openness concerns the results. In rational processes, often a timeframe is set. Processes of stakeholder involvement have their own dynamics, which makes it fairly difficult to predict what results will be delivered at a certain moment. When the process is fixed in time, the results that will be delivered should be indicated and not defined in terms that are too exact. To create more certainty, the process requires 'rules of the game' that contain rules for entering the process in later stages, how decisions are made, how information is brought into the process, etc. These 'rules of the game' should be discussed and should be approved by the stakeholders involved.

7. Tools, processes and instruments

As described in the previous section, the scientific literature usually distinguishes between five levels of stakeholder involvement, ranging from informing the stakeholders to making the stakeholders co-deciders. With these different levels of stakeholder involvement come different approaches in actually involving them. It would be impossible within the context of this section to describe all these different methodologies of stakeholder involvement. In Table 2 an overview is presented of possible tools, processes and instruments that can be used in the different levels of stakeholder involvement. Some methods are left out from Table 2 because they are not specific for the degree of influence of the stakeholders. These methods can be used in any process with some form of stakeholder participation: surveys, interviews, panel-research, idea- and complain-feedback forms, observations and hearings. For an overview, see the Consensus Building Handbook [23].

7.1. Case-study: dredging an artificial lake in Rotterdam, the Netherlands

Because of the need for sediment remediation and the improvement of water quality, a large lake near the city centre of Rotterdam had to be dredged. This meant that 175,000 m^3 of sediment had to be taken out and 100,000 m^3 of clean sand was needed to replace it. The total costs were 18 million euros. The area surrounding the lake is densely populated and, therefore, there was a requirement that the work should not become too much of a nuisance for the local citizens. Therefore, the project organization looked upon a number of challenges if they wanted to succeed. The areas that had to be used by the dredging company, such as islands and the shores of the lake, are owned by a diversity of residents, associations and companies. A possible threat to the success of the project could be the non-cooperation or even the obstructive power of these stakeholders.

The communication with the stakeholders was, therefore, taken very seriously. In the preparation phase of the project, clear communication with the stakeholders was organized. This consisted of the distribution of newsletters and the organization of information meetings, where the project approach was presented and the stakeholders could give input to improve and adjust the presented plans based on site-specific knowledge. The fact that they were taken seriously by the project organization also gave a firm support to the project, according to the representatives of a stakeholder group. In addition, because the project organization was clear and honest in their communication, it contributed to the support of the stakeholders.

Table 2. Stakeholder involvement and tools (source: adapted from [13, 20, 25])

Degrees of influence according to the scale	Role of the stakeholder	Role of the expert	Role of the policy-maker	Possible tools, processes and instruments to be used
1. Stakeholders are informed – they remain passive	Stakeholders receive information but do not deliver input to the process	Delivers information to the stakeholders on demand of the policy-makers	Policy-makers determine policy; information is issued to the stakeholders	Folders, brochures, leaflets, newsletters, advertisement, commercials, reports, exhibitions
2. Stakeholders are consulted	Stakeholders are consulted, act as interlocutors	Delivers information to the participants on demand of all parties; experts provide another flow of information to the process, next to the flow of the stakeholders	Policy-makers determine the policy and opens the process to input by stakeholders, but is not obliged to adopt their recommendations	Creative group sessions, study groups, focus groups
3. Stakeholders give advice	Stakeholders become advisors to the process	Delivers information to all parties on demand of all parties and investigate suggestions from participants on demand of the policy-makers	Policy process is open to input (other ideas, suggestions, etc.) by stakeholders; they take the input into account, but have the right to deviate from it in their decisions	Creative group sessions, advisory boards consisting of stakeholders, Internet discussion

Table 2 – *continued.*

4. Stakeholders become co-producers	Co-decision-makers within the set of preconditions	Experts treat policy-makers and stakeholders as equal clients; advice and knowledge provision to both actors	Policy-makers take the input of stakeholders into account, and honour it if it fits into the set of preconditions	Creative group sessions, project group where stakeholders also take part in producing solutions, Internet discussions
	Policy-partners on the basis of equivalence	Experts treat stakeholders as equal knowledge providers; they need approval of the stakeholders	Policy-makers interact with stakeholders on the basis of equivalence	Organizing workshops, create a common ground for discussion, for example by joint fact-finding
5. Stakeholders not only produce solutions but also decide about them	Taking initiatives, making decisions	Experts support stakeholders with knowledge; experts treat stakeholders as their clients, need no approval of the policy-makers	Joint role of policy-makers and actors: offer support (money, time of civil servants, etc.) and leaves the production of solutions and decisions to the participants	Joint groups that decide about implementation of solutions

When the project actually started, the communication continued and a special 'question and complain' phone line was created, together with so-called 'walk-in' mornings at the dredging site itself. According to stakeholders, they saw the clear and serious communication as the main reason why they were so supportive of the project. The project is generally seen as a success, which is partly due to good communication, the involvement of the stakeholders and the minimization of nuisance caused by the project to the stakeholders [31].

8. Risks and pitfalls

For stakeholder involvement, several risks and pitfalls are attached to these kinds of processes. *Asymmetry* in stakeholder involvement exists when some parties have an advantage over other parties. With asymmetry, there is a risk that the party who does not share a certain advantage may be overruled. At the same time, all parties usually have some kind of advantage, but not in the same area. Therefore, the challenge is to design the process in such a way that the different advantages are mixed and a mutual advantage will rise. This is something different than the rash conclusion that all parties should be equal in the process, as the process would not benefit from it. Some research on negotiation even points out that asymmetry is vital for the progress of the process. Perfect symmetry could result in a deadlock [32]. The existing asymmetries are an important factor for the design of the process, as they could be an important driving force, and therefore should be known when designing the process.

Asymmetries in stakeholder involvement not only include existing advantages, but also lack of representation of stakeholders, the knowledge gap (as not all stakeholders have the same level of knowledge), different interests and the lack of communication [20]. Lack of representation means that stakeholders are not representing the target group. In some cases this leads to 'extreme views' that are not representative of the opinion of the target group. The full range of perspectives and interests are not taken into account.

The *knowledge gap* or information asymmetry follows from the observation that not all stakeholders are equal and poses different information and knowledge. The knowledge gap is of particular concern with the issue of sediment. Sediment issues are highly specific, and often require sophisticated knowledge to understand them. Even among experts there is still considerable debate concerning the understanding of, for example, morphology. At the same time, experts may lack knowledge as well. This is not a disqualification; it just follows from the fact that not all participants are equal. Too often, the knowledge gap is regarded as the lack of knowledge with laymen. This indeed is often the case and calls for a flow of information to the other participants and the development of a common ground of knowledge. But at the same time, laymen have other knowledge (i.e. 'lay-knowledge', information about the local situation, etc.) at their disposal. This is as valuable as scientific knowledge and should not be ignored. A final remark on the knowledge gap is that knowledge can be used as a weapon to get one's point of view accepted.

Connected to this is the pitfall of *lack of communication*. The language used by experts is very different and often incomprehensible to laymen. This can create confusion and distrust as other participants might feel that the party concerned holds back information.

Another type of pitfall that comes under the heading of 'asymmetry' is the different interests and needs of participants. Stakeholders all have different agendas and a pitfall is ignoring some of them or assuming that everyone is aiming at the same goal. This not only applies to individuals but also to countries. Western Europe might be concerned about the environmental impact of polluted sediment, whereas developing countries are usually more concerned about financial considerations.

Clashing expectations often exist as participants have different expectations and consequently expect different outcomes. 'For example, a governing body of a river can invite people living near a dredged material dumpsite to come up with new ideas about how to address the dumping of contaminated sediments. They are consulted and asked to give a recommendation. However, should this not be properly communicated, then the invitees might believe that they are expected to take part in the decision-making. The result will be that their expectations rise too high, and thus cannot be met, resulting in distrust, downright pessimism and obstruction of the process' [20].

Stakeholder out-of-sight situations often exist in the formal decision-making process. Unfortunately, a sharp separation is made between the stakeholder process and the actual decision-making. The process of stakeholder involvement is then regarded as a way to pacify the opposition, where the actual decision mainly serves the interests of the formal decision-maker. Decision-makers should commit themselves to the process, whatever the outcomes.

Cross-boundary cooperation presents a whole range of pitfalls and challenges (see Chapter 8, this book). The management of sediment at the river basin scale will often be cross-boundary and, therefore, attention must be paid to the characteristics of international cooperation in stakeholder processes, especially where the cultural differences deserve attention. Earlier in this chapter we discussed the different perspectives from which people look at things. Cultural differences can be regarded in the same way because they are closely connected to parties and determine their way of thinking and behaving [33]. The interpretation of different cultures is not easy and requires time [34]. Apart from that, different countries have different institutions. Although countries may be adjacent to each other, they still can differ a lot which can create confusion and fragmentation of the decision-making process [35]. Legislation is also part of the institutional dimension. In both national and international legislation, laws and regulations concerning river basin management are fragmented ([36]; Chapter 3, this book). For the process of stakeholder involvement, this means that the participants should be given time to frame the process into their own legislation. Otherwise, the proposals rising from the process will be lost in the misfit of legislation.

9. Conclusions

This chapter provided an introduction to the subject of stakeholder involvement with respect to sediment management. The relevant themes have been grouped around the basic questions, namely who, what, why and how? Furthermore, the pitfalls of stakeholder involvement were mentioned as the process of stakeholder involvement is not as simple as it may seem at first glance. The need for stakeholder involvement in sediment management issues is a clear one. Dealing with sediment management at the river basin scale is a complex policy issue with a wide variety of different policy levels (see Chapter 3, this book) and stakeholders involved, with different interests and perspectives. Future management of sediment – whether at specific sites or at the river basin scale – will have to incorporate the views, interests and perspectives of the various stakeholders. This is not an easy task, because of the lack of a 'common language' and mutual understanding is not simple. The stakeholder process, therefore, deserves a lot of attention and it should be done in a serious way, whereas people that are not taken seriously will be disappointed and pull out of the process. The general unawareness of the general public and the complexity of the sediment issue are important hurdles to overcome. This subject deserves much attention in the policy process and stakeholder involvement. Experts should focus on making their expert knowledge available to the general public and take time to explain the issue. For further information on the role of stakeholders in sediment management at the river basin scale, including specific examples, see Joziasse *et al.* [37], Ellen *et al.* [38] and Chapter 8 (this book).

References

1. Renn O, Webler T, Wiedemann P (1995). Fairness and Competence in Citizen Participation. Evaluating Models for Environmental Discourse. Kluwer Academic, Dordrecht.
2. Healey P (1997). Collaborative Planning. Shaping Places in Fragmented Societies. Macmillan, London.
3. Coenen F, Huitema D, O'Toole Jr L (Eds) (1998). Participation and the Quality of Environmental Decision Making. Kluwer Academic, Dordrecht.
4. Tunstall S, Tapsell S, Eden S (1999). How stable are public responses to changing local environments? A 'before' and 'after' case study of river restoration. Journal of Environmental Planning and Management, 42: 527–547.
5. DeLeon P (1992). The democratization of the policy sciences. Public Administration Review, 52: 125–129.

6. DeLeon P (1994). Democracy and the policy sciences. Aspirations and operations. Policy Studies Journal, 22: 200–212.
7. Durning D (1993). Participatory policy analysis in a social service agency: a case study. Journal of Policy Analysis and Management, 12: 297–322.
8. Fischer F (2000). Citizens, Experts and the Environment. The Politics of Local Knowledge. Duke University Press, Durham, NC.
9. Mason M (2000). Evaluating participative capacity-building in environmental policy: provincial fish protection and parks management in British Columbia, Canada. Policy Studies, 21: 77–98.
10. Dobbs L, Moore C (2002). Engaging communities in area-based regeneration: the role of participatory evaluation. Policy Studies, 23: 157–171.
11. Murray M, Greer J (2002). Participatory planning as dialogue: the Northern Ireland Regional Strategic Framework and its public examination process. Policy Studies, 23: 191–209.
12. Verweij M, Josling T (2003). Deliberately democratizing multilateral organization. Governance, 16: 1–21.
13. Edelenbos J (2000). Process in Shape. Lemma, Utrecht.
14. Hahn T (2003). Paper presented at the Greening of the Industry Network Conference 12–15 October, San Francisco, CA.
15. Freeman RE (1984). Strategic Management: A Stakeholder Approach. Pitman, Boston, MA.
16. Mitchell R, Agle B, Wood DJ (1997). Toward a theory of stakeholder identification and salience: defining the principle of who or what really counts. Academy of Management Review, 22: 853–886.
17. Donaldson T, Preston L (1995). Stakeholder theory of the corporation: concepts, evidence, implications. Academy of Management Review, 20: 65–91.
18. Clarkson M (1995). A stakeholder framework for analyzing and evaluating corporate social performance. Academy of Management Review, 20: 92–117.
19. Rowley T (1997). Moving beyond dyadic ties: a network theory of stakeholder influences. Academy of Management Review, 22: 887–910.
20. Gerrits L, Edelenbos J (2004). Management of sediments through stakeholder involvement. Journal of Soils and Sediments, 4: 239–246.
21. Norris P (Ed.) (1998). Critical Citizens. Global Support for Democratic Government. Oxford University Press, Oxford.
22. van Ast JA (2000). Interactive Water Management in Cross-Boundary River Systems. Eburon, Delft.
23. Susskind L (Ed.) (1999). The Consensus Building Handbook: A Comprehensive Guide to Reaching Agreement. Sage, Thousand Oaks, CA.
24. National Research Council (2001). A Risk-Management Strategy for PCB-Contaminated Sediments. National Academy Press, Washington, DC.
25. Pröpper IMAM, Steenbeek D (1999). De Aanpak van Interactief Beleid: Elke Situatie is Anders. Coutinho, Bussum.
26. Thompson M, Ellis R, Wildavsky A (1990). Cultural Theory. Political Cultures Series. Westview Press, Boulder, CO.
27. Schön DA, Rein M (1994). Frame Reflection: Toward the Resolution of Intractable Policy Controversies. Basic Books, New York.
28. van Asselt MBA (2000). Perspectives on Certainty and Risk. The PRIMA Approach to Decision Support. Kluwer Academic, Dordrecht.
29. van de Riet O (2003). Policy Analysis in Multi-actor Policy Settings, Navigating between Negotiated Nonsense and Superfluous Knowledge. Eburon, Delft.

30. Ellen GJ, Slob AFL (2005). Risk perception and risk communication concerning sediments. In: Olfenbuttel RF, White PJ (Eds), Remediation of Contaminated Sediments – 2005: Finding Achievable Risk Reduction Solutions. Proceedings of the Third International Conference on Remediation of Contaminated Sediments (New Orleans, Louisiana, 24–27 January 2005). Battelle Press, Columbus, OH, pp. 1–20.
31. Hermans B (2005). Stilstand is Achteruitgang. Kansen Benutten van de Europese Kaderrichtlijn Water. Stichting Natuur en Milieu and the Provinciale Milieufederaties, Utrecht.
32. Zartman IW, Rubin JZ (2003). Power and Negotiation. The University of Michigan and the International Institute for Applied Systems Analysis, Michigan.
33. Hall ET, Hall E (1976). How cultures collide. Psychology Today, 10: 66–74.
34. Sperber D (1996). Explaining Culture: A Naturalistic Approach. Blackwell, Oxford.
35. de Jong M (1999). Institutional Transplantation. How to Adopt Good Transport Infrastructure Decision-Making Ideas from Other Countries? Eburon, Delft.
36. Palmer CG, Peckham B, Soltau F (2000). The role of legislation in river conservation. In: Boon PJ, Davies BR, Petts GE (Eds), Global Perspective on River Conservation. John Wiley and Sons, Chichester, pp. 474–491.
37. Joziasse J, Heise S, Oen A, Ellen GJ, Gerrits L (2007). Sediment management objectives and risks indicators. In: Heise S (Ed.), Sustainable Management of Sediment Resources: Sediment Risk Management and Communication. Elsevier, Amsterdam, pp. 9–75.
38. Ellen GJ, Gerrits L, Slob AFL (2007). Risk perception and risk communication. In: Heise S (Ed.), Sustainable Management of Sediment Resources: Sediment Risk Management and Communication. Elsevier, Amsterdam, pp. 233–247.

Sustainable Management of Sediment Resources: Sediment Management at the River Basin Scale
Edited by Philip N. Owens

Towards Sustainable Sediment Management at the River Basin Scale

Philip N. Owens[a], Adriaan F.L. Slob[b], Igor Liska[c] and Jos Brils[d]

[a]*National Soil Resources Institute, Cranfield University, North Wyke Research Station, Okehampton, Devon EX20 2SB, UK. Now at: Environmental Science Program, University of Northern British Columbia, 3333 University Way, Prince George, British Columbia, V2N 4Z9, Canada*
[b]*TNO Built Environment and Geosciences, PO Box 49, 2600 AA Delft, The Netherlands*
[c]*International Commission for the Protection of the Danube River, Vienna, Austria*
[d]*TNO Built Environment and Geosciences, PO Box 80015, 3508 TA Utrecht, The Netherlands*

1. Introduction

The previous chapters of this book have presented information on some of the key requirements for decision-making for the management of sediment at the river basin scale. These include:

- identifying the many functions that sediment has within river basins, understanding the behavioural characteristics of sediment that enable it to perform these functions, and recognizing that the river basin represents the most appropriate unit for management decision-making so as to maintain, or influence, these functions (Chapter 1);
- developing conceptual frameworks for sediment management that address both site-specific and basin-scale needs, which in turn form part of a larger basin management plan, and incorporate many of the requirements listed below (Chapter 2);
- understanding and assessing the relevant legislative and non-legislative drivers that influence why and how sediment is, or should be, managed (Chapter 3);
- understanding the movement of sediment and associated contaminants in terms of the sources and transport pathways of sediments and contaminants throughout the river basin, and how these respond to natural and anthropogenic changes including sediment management (Chapter 4);

- using tools to assemble the information required to understand the sediment–contaminant system, both now and in the future (Chapter 5), and to assist in balancing the costs and benefits of management options (Chapter 6); and
- identifying stakeholders and involving them in the decision-making process (Chapter 7).

Other specific requirements are addressed in the books on sediment quality and impact assessment of pollutants by Barceló and Petrovic [1], sediment and dredged material treatment by Bortone and Palumbo [2], and sediment risk management and communication by Heise [3].

The purpose of this chapter is to place these requirements for, and approaches to, sediment management within the broader historical context of environmental sustainability (Section 2). Understanding this historical development of environmental policies is important because it helps to place sediment management within a broader context of environmental management, and to understand where such programmes may go in the future. This chapter also discusses some of the options available for sustainable sediment management, and provides some critical comments on what is required (Section 3). Section 4 presents the steps that should be considered within an adaptive framework for river basin management. Section 5 presents examples of recent initiatives for improved sediment management within Europe, including national and multinational scientific research projects and collaborative initiatives that aim to encourage the assemblage and communication of data and ideas on sediment within river basin management. Finally, some conclusions and recommendations are put forward (Section 6).

2. The use of science in sediment management at a river basin scale

The different chapters in this book look at sediment management at the river basin scale. This leads to an increasing complexity, because now a system approach should be followed addressing the river–sediment–soil–groundwater system. It brings sediment management into the context of river basin management and hence into contact with a variety of related issues (such as nature conservation, the use of space, economic and social issues) and stakeholders (with concerns ranging from nature conservation to entrepreneurial interests) who want to be involved in river basin management. Because of this development, a shift is taking place in the ways that science, scientific knowledge and technical expertise will be used in sediment management. The technical, monodisciplinary approach to sediment management, which has been common practice until now, does not fit well with these developments because it neglects the complexity of the river basin system. A new management and policy approach, as well as new ways of involving science in policy, are needed. This section places these developments within a historical perspective and explores the new role of science in managing sediment in the context of river basin management.

2.1. Four generations of environmental policies

Environmental policies have developed continuously from the 1960s, when environmental problems were first recognized, and various authors (e.g. [4–6]) have described this development. For example, four generations of environmental policies can be distinguished that were developed in the Netherlands [7]. This development, although specific for environmental policies and specific to the Netherlands, can also be recognized in other policy areas and other countries (see Chapter 3, this book), and thus serves as a useful example. In different policy areas there is a trend of increasing complexity, which is expressed by the growing number of stakeholders that are involved and by the connection to other policy areas.

The first generation of environmental policies started in the 1960s and 1970s. Because of poor environmental conditions and health problems in some regions, an environmental consciousness awoke in the public and the government adopted new sets of environmental regulations and rules. For the most part, these environmental measures involved end-of-pipe techniques to reduce emissions. At the beginning of the 1980s, awareness increased that the end-of-pipe-measures on their own were not enough to tackle adequately the environmental problems of the time. This cleared the way for the prevention of environmental damage. The second generation of environmental policies focused strongly on prevention by a mix of voluntary measures and regulations. The focus of these voluntary and regulatory measures was, for example, on energy saving and waste reduction. The environment had become a topic in quality and safety systems of businesses and environmental management systems had become popular. So in the second generation of environmental policies, the attention shifted from end-of-pipe techniques to process-integrated measures designed to prevent pollution being emitted to the environment. This second generation of environmental policies developed further in the 1990s, when the attention shifted from prevention within the factory to prevention of emissions within the whole production–consumption chain. Product-oriented policies and chain management are typical examples of the third generation of environmental policies. The environmental aspects of products were measured during their entire life cycle and incorporated in chain management and the product design process. Major efforts were put into communication with external stakeholders concerning environmental performance. The policy instruments that were being used were of a voluntary nature, sometimes supported by legislative and/or financial measures (see Chapter 3, this book). At the beginning of the 21st century, new strategies were needed because persistent environmental problems were still present. Some problems, e.g. acidification, had been tackled quite well, but more persistent problems, e.g. the emission of CO_2 or the spreading of chemicals in the environment, were harder to tackle. So at the beginning of this century the need for a fourth generation of environmental policies was articulated. Although this generation of environmental policies is still developing, we can already see its early stages. For example, in the last Dutch Environmental Policy Plan 4 (2001) [8] 'transition management' is presented as a key concept to tackle persistent environmental problems. The policy approach is process-oriented instead of content-oriented and leans on societal dialogue and collaborative processes.

In Table 1, an overview of the four generations of environmental policies is presented. When the characteristics of the generations are examined more closely, a clear trend becomes apparent. There is an increase in complexity of the environmental problems, which is reflected in an increase in the number of stakeholders involved, distributed responsibilities and knowledge, and a very long time horizon. Whereas in the first generation, the goals that were set were quite straightforward, legislative instruments were used and the number of actors was limited, in the fourth generation, goals are discussed and formed with societal actors over a long period, participative approaches are important and the number of actors is numerous. Whereas the policies in the first generation were directly aimed at cleaning the environment in a fairly straightforward way, in the fourth generation this aim has shifted towards upgrading environmental quality by means of system innovations and societal change. A main driver in the fourth generation is 'sustainability', more formally known as 'sustainable development'. In the next subsection this concept will be explored.

2.2. The concept of sustainable development

The concept of sustainable development was first introduced in 'Our Common Future' [9], also known as the Brundtland Report, which was a milestone for environmental policy and drew much attention when it was published in 1987. This concept became an important policy principle for many governments worldwide. It brought the environmental issue into the mainstream of social and economic development. In 'Our Common Future', sustainable development is defined as a 'development that meets the needs of the present without compromising the ability of future generations to meet their own needs'. Sustainable development has widened the scope of environmental policies to other policy domains and has linked environmental thinking to development issues, especially to social and economic policy issues. In Figure 1 these major

Table 1. Characteristics of four generations of environmental policies

	First generation	Second generation	Third generation	Fourth generation
Means to reduce environmental damage	Clean-up operations	Prevention	Chain management	Network management
Policy instruments	Legislation and regulation	Regulation, voluntary measures	Voluntary measures, financial instruments, regulations	Participatory instruments
Scope	Substances, emissions	Processes	Products, production chain processes	Sustainability, societal processes, system innovations
Time span	Some years	One year	Some years	More than a generation (20 years)
Central actor(s)	Government	Government, management of a company	Government, management of different companies, consumer organizations	Government, companies, stakeholder groups, etc.
Actors involved	Stakeholders	Employees of a company	Companies, stakeholders, consumers	Societal groups, stakeholders, etc.
Drivers for actors	Legislation and external pressure	Efficiency	Strategic performance	Sustainability, licence to operate
Actions	New technology, registration, monitoring	Process changes, communication (internal)	Product (re)design, balanced scorecard, covenants, external communication	Societal dialogue, institutional change, societal reform
Complexity	Low	Moderate	High	Very high

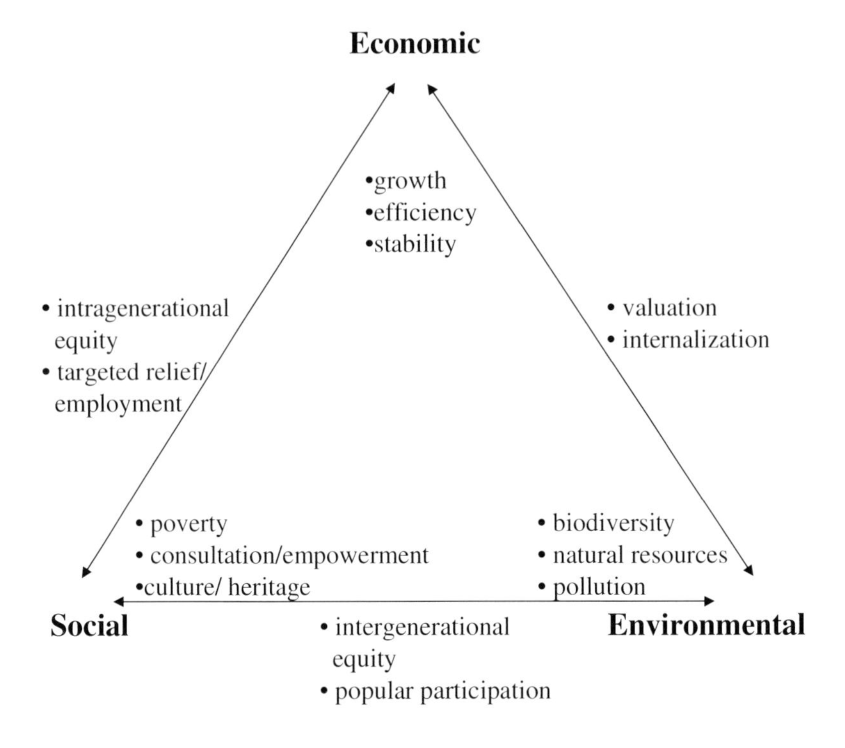

Figure 1. The three elements of sustainable development and their interactions (source: modified from [10])

elements of sustainable development – economic, social and environmental – are shown [10].

In the corners of the triangle are the important issues of the three elements of sustainable development:

- economic: growth, efficiency, stability;
- social: poverty, consultation/empowerment, culture/heritage; and
- environmental: biodiversity/resilience, natural resources, pollution.

The issues that interact with two elements are placed on the sides of the triangle:

- social–economic interface: intragenerational equity (income distribution); targeted relief of the poor and employment;
- social–environmental interface: intergenerational equity (rights of future generations), public participation; and
- economic–environmental interface: valuation, internalization.

Achieving sustainable development is a formidable task, since all three elements must be given balanced consideration [10]. Nevertheless, Figure 1 shows which issues must be addressed in giving direction to sustainable development. An important premise in the Brundtland Report is that interventions can direct sustainable development. The belief of sustainable development as a process of directed change of social evolution is embedded heavily in the report: 'Humanity has the ability to make development sustainable. ... The concept of sustainable development does imply limits – not absolute limits but limitations by the present state of technology and social organization on environmental resources and by the ability of the biosphere to absorb the effects of human activities. But technology and social organization can be both managed and improved to make way for a new era of economic growth' [9]. This emphasizes the role of technology and social organization in achieving sustainable development, i.e. the role of innovation in both technology and social organization. Sustainable development implies a process of social learning, for instance it is about sharing visions on what *has* to be sustained. In such a process of social and mutual learning, interaction between

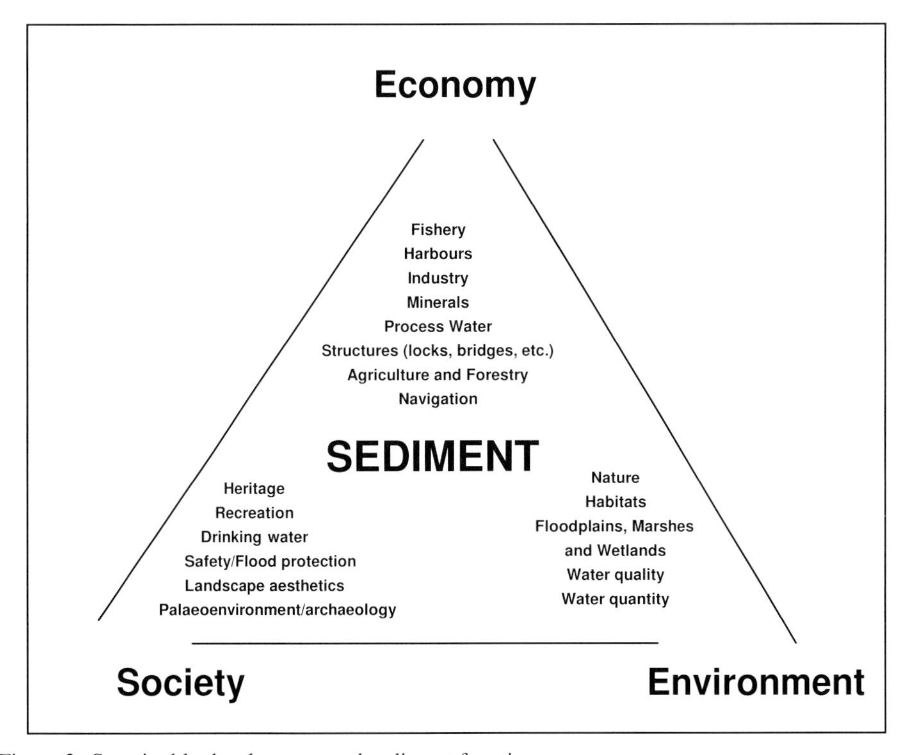

Figure 2. Sustainable development and sediment functions

the actors involved is a necessary element. In this interaction, the balance between the three 'corners' of sustainable development has to be found on a regional or local level in a certain time frame. Embracing the concept of sustainable management in policy-making should therefore lead to involvement of different stakeholders in an early stage of policy-making (see Chapter 7, this book).

The concept of sustainable development can be applied specifically to sediment issues. Sediment has several functions that play an essential role in the river and marine systems, and must be viewed in this context. For a description of these functions see Chapter 1 (this book), Salomons and Brils [11], and Owens *et al.* [12]. In Figure 2, several functions of sediment are related to, and categorized along, the three 'corners' of sustainable development.

2.3. Science in the fourth generation

In the previous sections the development of environmental policies was explored in depth. We can now deduce what this might mean for knowledge production in the fourth generation of environmental policy. In the first generation, the demand for knowledge comes only from government and is quite technical in nature. In the third and fourth generations, the actors are quite numerous and the demand for knowledge comes now from different actor groups. Therefore, the new policy approach in the third and, especially, the fourth generation will call for a new way of dealing with knowledge in the policy process. This fits well with the ideas of Gibbons *et al.* [13], who reflected on the role of science in society. According to them, science is undergoing a major shift from mode 1 science – the traditional way of production of scientific knowledge – to mode 2 science. In this mode 2 approach, the societal context is very important for knowledge production. Mode 2 science has five characteristics that distinguish it from mode 1 science, and these are listed in Table 2.

Table 2. Characteristics of mode 1 and mode 2 science (source: [13])

	Mode 1	Mode 2
Knowledge developed in:	Academic context	Context of application
Knowledge production:	(Mono)disciplinary	Transdisciplinary
Place and way of knowledge production:	Homogeneous (one place, a certain time)	Heterogeneous (knowledge developed close to the place of application)
Organization:	Hierarchical, preserves form	Heterarchical, transient
Quality control:	Academic, peer review	Socially accountable, reflexive

Mode 2 science is generated in the context of application, whereas mode 1 science is produced in the 'classic' academic environment. 'The context of application describes the total environment in which scientific problems arise, methodologies are developed, outcomes are disseminated, and uses are defined' [14]. Furthermore, mode 2 science is produced in a 'transdisciplinary' way, 'by which is meant the mobilization of a range of theoretical perspectives and practical methodologies to solve problems. But, unlike inter- or multi-disciplinarity, it is not necessarily derived from pre-existing disciplines nor does it always contribute to the formation of new disciplines' [14]. Mode 1 science is produced in the academic setting (one place, a certain time), is organized in a hierarchical way and involves academic peers to control the quality of the output. Mode 2 science is developed close to the place of application, is heterarchically organized and quality control is performed by societal actors. Mode 2 science fits quite well to the fourth generation of environmental policy because of the social accountability, the fact that it recognizes the importance of stakeholder views, and, more generally, because of the emphasis on the context of application.

From this concept, we can deduce some key elements for the new ways of knowledge production. Knowledge production should be socially accountable and should acknowledge the multiple rationalities and different viewpoints that are brought in by the variety of stakeholders that are involved (see Chapter 7, this book). The research process should contain the following key elements:

- Multi- or transdisciplinary research methods. In research many disciplines should be involved to give more insight for the stakeholders in the policy issues, the framing of problems and the development of solutions. This means that the different disciplines should work together from the very start and that common concepts should be developed.
- Involvement of stakeholders in the research process. Methods to involve stakeholders and to deal with their viewpoints and values in the research process should be developed. Appreciation of the values, interests and viewpoints of involved stakeholders requires the creation of new research methods to develop and share knowledge. The instrument of 'joint fact finding' [15] is already available for this (see also Chapter 6, this book).
- Emphasis on learning approaches in management and policy processes. Learning between stakeholders, between scientists, between stakeholders and scientists, on different levels (individual, team and organization) has a central role in the fourth generation of environmental policies and requires special arrangements. Organization of reflection, feedback and evaluation of the goal achievement of the policy is needed, and monitoring plays a key role in that.

Based on the recent development of environmental policy, and in particular the development of environmental sustainability, and how this influences why and how we manage environmental resources, the following sections describe the approaches and requirements for sustainable management of sediment.

3. Towards sustainable management of sediment

3.1. Linking basin to local scales: breaking down and building up

Today, river management is moving towards sustainability, working with natural processes at the basin scale [16]. As Section 2 has identified, environmental concerns are widely recognized as drivers for basin management and local, site-specific issues should be considered within the context of the wider environmental landscape, of which the river basin is an appropriate unit and scale for water and sediment [3, 16, 17]. The virtues of the basin-scale approach to land–water management have been extolled by the chapters in this book, as well as many of those in Heise [3] and elsewhere. However, 'the characteristics and processes involved are typically too complex in a river catchment to be immediately understood as an entire entity' [16] or system. A more realistic, and ultimately more useful, approach is to separate the basin into manageable components or units that can be studied before using these components to build up towards an understanding of the whole system or basin [16, 17]. This breaking down–building up approach to river basin management is illustrated in Figure 3. Key within this approach is the fact that the river basin often represents the start- and end-point of the process, with any local or site-specific issues nested within this. This approach is largely consistent with many of the strategic and conceptual frameworks described in Chapter 2 (this book) (see also [18]), where River Basin Management Plans are the goal, with process understanding and risk assessment being part of the process to achieve this.

To some extent, the breaking down–building up approach to river management has been going on for decades, although mainly by each component or sub-system being addressed by a specific discipline or sub-discipline, such as hydrology, geomorphology, ecology, engineering, economics and sociology. What is becoming increasingly common, and is required for future progress in this field, is the integration of these disciplines as river basins are both broken down into understandable and/or measurable units, and are then reassembled to provide an integrated understanding of the whole basin and how it functions as such. Section 5 provides case study examples of where progress is being made using a more integrated approach that is able to link different spatial and temporal scales of interest, and also link physical, economic and

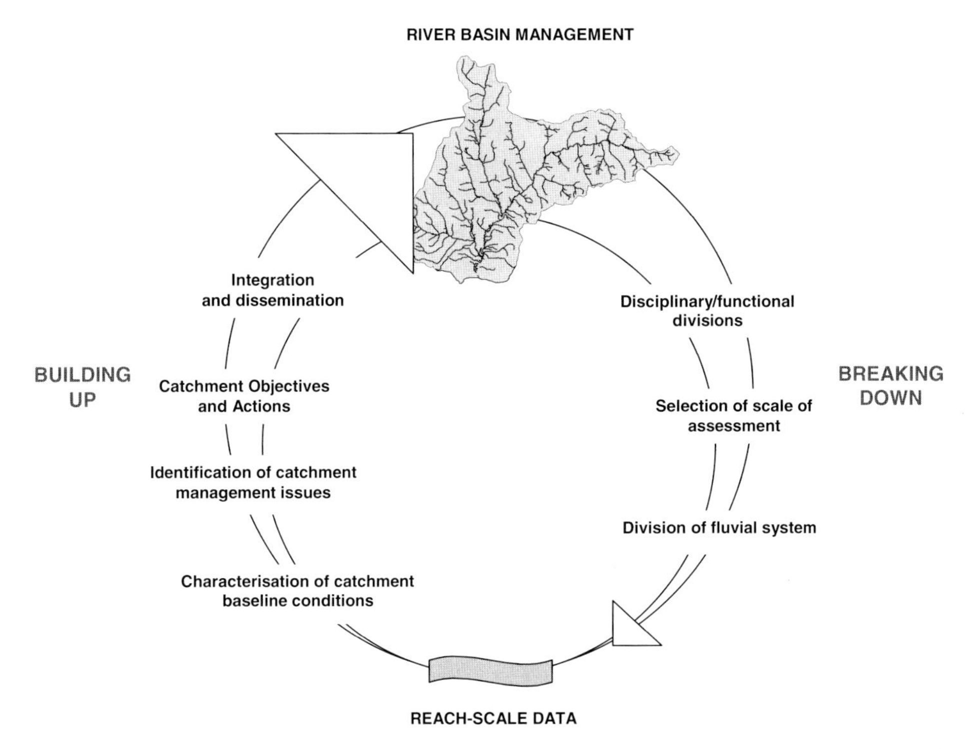

Figure 3. Breaking down and building up approach to river basin management (source: [16] reproduced with the permission of Blackwell)

social requirements. In addition, such an approach sits well within the fourth generation of environmental policies described in Section 2.

While the approach described above and illustrated in Figure 3 offers a way to link river basin and local scales (also see [3], Chapter 2, this book), how we address issues of uncertainty, and the role of emergent properties, as we move between scales is important. There is a certain degree of uncertainty with almost all pieces of information. Estimates of sediment fluxes, for example, have uncertainties due to errors in field sampling/measurement and laboratory analysis ([19, 20], Chapters 4 and 5, this book). These errors are often compounded within numerical models when one data set is used with other data types, each of which has its own errors and uncertainties (see Chapter 5, this book). Similarly, information from stakeholders, such as opinions, historical information etc., have uncertainties that can alter as we move between local and basin scales, and that can influence decisions. These are issues that require further understanding.

3.2. Sustainable solutions that operate at the basin scale

In order to achieve more sustainable use of river basins, there is a need to develop solutions that operate at the basin scale and that address the numerous environmental, economical and societal functions that sediment performs within basins (see Figure 2). In this respect, the control of the *sources* of sediment and associated contaminants represents perhaps the best long-term solution. It may also represent one of the most cost-effective solutions in many cases because the solution targets the cause and not just the effect. Dredging represents a good example. Dealing with dredged sediment is a costly management issue, especially if the dredged material is deemed to be contaminated. The Port of Rotterdam in The Netherlands, for example, dredges about 20×10^6 m^3 of sediment annually, of which >90% is not contaminated and can be relocated at sea [21]. In the case of the Port of Hamburg, Germany, about $3–5\times10^6$ m^3 $year^{-1}$ of sediment is dredged, much of which is contaminated [22], while the volume of dredged marine sediments obstructing ports in France is ca. 50×10^6 m^3 $year^{-1}$ [23]. It is estimated that the costs of dealing with the sediment that is dredged from the Port of Hamburg and treated is of the order of 30 million euros per year [24]. The book by Bortone and Palumbo [2] provides a good review of treatment approaches and costs. However, if the supply of sediment and contaminants is controlled through the identification of sediment–contaminant sources (see Chapters 4 and 5, this book), and appropriate remediation measures are put in place, this deals with the problem at the origin and not further down the system. If, for example, the source of the sediment is due to soil erosion on a certain type of agricultural land, such as intensive arable land, then the solution can be a win-win one in that both downstream dredging costs and on-field soil loss costs (e.g. reduction of the loss of agriculturally productive soil) can be reduced. Of course, here the use of (societal) cost-benefit analysis is important, in order to evaluate the costs of implementing a solution relative to the benefits gained or relative to the costs of achieving the same benefits using a different solution (see also Chapter 6, this book).

While the principle of source control for sustainable river basin management may appear to be an obvious one, there are some potential obstacles to overcome. One of these is identifying the source of the sediment and/or contaminants. Chapters 4 and 5 (this book) have provided examples of methods for identifying sediment sources, and presented case studies to illustrate the type of information that can be obtained (e.g. Table 1 and Figure 4 in Chapter 4; Figures 3 and 5 in Chapter 5). But it must be recognized that such information can itself be costly to assemble and that there are uncertainties associated with the results.

Another problem is the implementation of a management strategy to actually control the mobilization and transport of sediments and contaminants once the source type or source area has been identified. In other words, identifying the sources may be relatively easy, but actually doing something about it may be more problematic and there may be many obstacles in the way to the successful uptake of a management plan. A good example is agriculture. In many river basins in Europe, and indeed the world, monitoring and tracing studies have demonstrated that much of the sediment (and some contaminants) will be derived from agricultural land (see Chapter 4, this book). Table 3 reviews the types, occurrences and severity of soil erosion at the national level in Europe, most of which is on agricultural land and a function of intensive agricultural practices.

Due to the widespread occurrence of agricultural land in the countries listed in Table 3, soil erosion on land and sediment delivery to rivers is best controlled by land management and the adoption of measures to control soil erosion and sediment transfers on land. Such solutions require regional, national or multi-national policies as incentives, or legislation to enforce change in land use and management at large spatial scales. In the EU, soil erosion on agricultural land and sediment delivery to rivers is partly a function of EU agricultural policies such as the Common Agricultural Policy (CAP) and CAP-Reform. There are some national policies and initiatives aimed at reducing sediment and contaminant delivery in river basins, especially from diffuse sources, such as the Catchment Sensitive Farming programme in the UK (see Chapter 3, this book) which operates at the catchment or river basin scale. Such regional, national or international agri-environment programmes, which are themselves usually part of a broader programme of agricultural and/or environmental policies, offer much potential for dealing with large-scale diffuse issues such as soil erosion, sediment delivery and diffuse pollution (also see [25]).

A promising approach for identifying and prioritizing actions with sediment–contaminant problems at the basin scale is that developed by Heise *et al.* [27] for the drainage basins of the Rivers Elbe and Rhine. The driver for this work is the deposition of sediment, some of which is contaminated, in the ports of Hamburg and Rotterdam, respectively. The approach links identifying the sources of concern within a risk approach. Heise *et al.* [27] apply a three-step approach in order to prioritize sites in the river basin with regard to the risk that they pose for downstream areas:

- identification of the substances of concern and classification of these into 'hazard classes of compounds';
- identification of areas of concern and their classification into 'hazard classes of sites'; and
- identification of areas of risk.

Table 3. Types, occurrence and severity of soil erosion at the national level in Europe. Numbers relate to the degree of erosion: 1 = minor, 2 = important; 3 = predominant, N = not found, ? = not known (source: modified from [26])

Country	Rill & interrill	Gully	Snow melt	Bank	Tillage	Animals	Wind	Land-slides
Albania	2	2	1	2	1	2	?	2
Austria	2	1	2	2	1	N	?	2
Belgium	2	1	N	1	1	N	1	1
Bosnia/Herzeg.[a]	2	2	1	2	1	1	1	1
Bulgaria	2	2	2	1	1	1	1	1
Croatia	2	2	1	2	2	1	1	2
Cyprus	2	2	1	1	2	?	?	1
Czech R.	3	1	1	1	1	?	?	1
Denmark	3	1	N	1	1	N	2	?
Estonia	2	N	N	?	?	1	1	N
Finland	1	N	2	1	?	1	N	N
France	3	2	2	3		1	1	2
Germany	2	1	1	2	1	?	2	2
Greece	1	3	1	3	1	2	1	1
Hungary	2	2	1	2	2	1	1	1
Iceland	1	2	3	3	N	N	1	2
Ireland	1	N	N	3	1	2	N	N
Italy	3	2	1	1	2	?	1	2
Latvia	2	N	N	?	?	1	?	N
Lithuania	2	N	N	?	?	1	?	N
Luxembourg	1	N	N	1	N	N	N	?
Macedonia	2	2	1	2	1	1	?	1
Malta	1	2	N	N	N	1	N	1
Montenegro	2	2	1	2	1	2	?	1
Norway	1	1	3	1	N	1	1	2
Poland	2	1	1	1	?	?	2	2
Portugal	2	3	N	1	1	?	?	1
Romania	2	2	1	2	1	1	?	1
Serbia	2	2	1	2	1	1	1	1
Slovakia	2	1	?	1	?	?	?	1
Slovenia	2	2	1	1	2	?	?	2
Spain	2	3	1	1	1	1	1	2
Sweden	1	2	1	2	N	1	1	3
Switzerland	1	1	3	2	?	1	?	2
The Netherlands	1	N	N	?	N	?	1	N
UK	2	1	1	2	1	2	1	1

[a]Herzegovina

Figure 4 illustrates how this approach has been used to identify and classify areas of risk based on historically contaminated sediment in the Rhine basin in Germany.

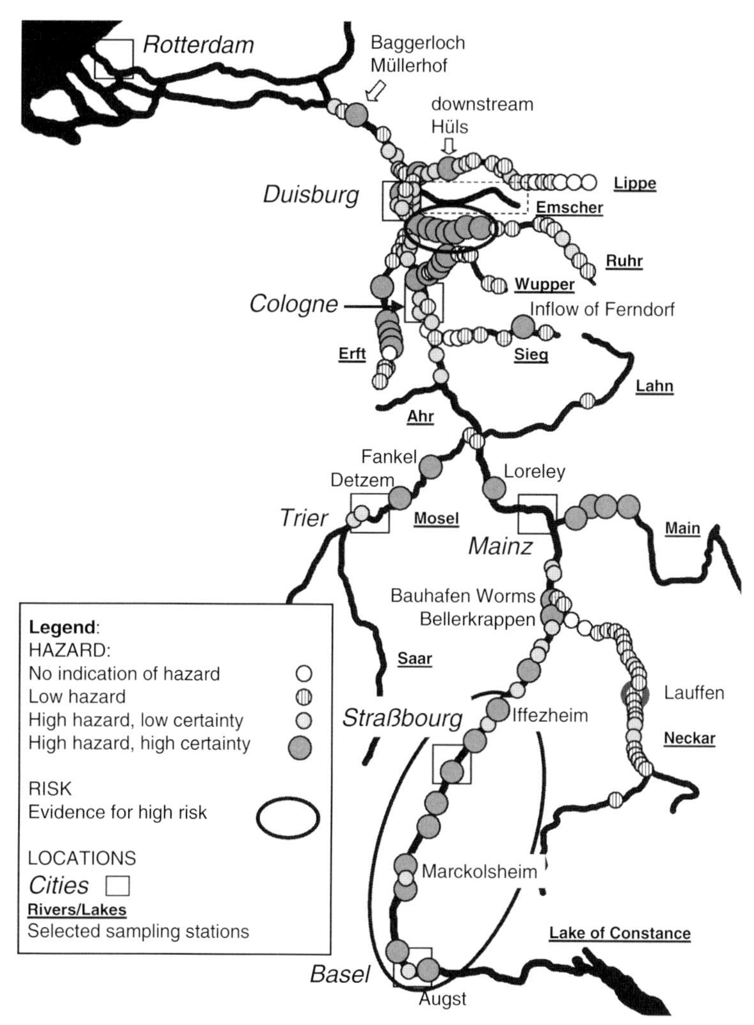

Figure 4. Sediment–contaminant risk assessment for the Rhine basin, Germany, based on the need to mitigate downstream sediment issues in Rotterdam Port (source: [27, 28] reproduced with the permission of Springer)

3.3. Addressing social, economic and environmental sediment issues within the sustainability concept

For sediment management to be truly sustainable, perhaps the most important thing that we need to address is how we balance the three elements of the sustainability concept described in Section 2: society; economy; and environment (see Figure 2). Most of the previous chapters have been concerned with one, or perhaps two, elements and we still have problems in incorporating all three elements into both our thinking and our actions. In part this stems from our limited capacity to understand all parts of the system (social, economic and environmental) and our inability to copy with areas outside of our experience and expertise. The concepts described in Section 2 and decision-support tools, such as societal cost-benefit analysis (Chapter 6, this book), offer ways to address simultaneously two or three elements of sustainable development. At the moment, we are not yet at the stage where we are able to incorporate all three elements of sustainability equally in the decision-making process. This book, for example, mainly focuses on environmental, and to some degree social and economic, considerations. Other books in this series [1–3] tend to focus mainly on environmental issues of sediment management. While not a criticism of these books, it helps to further illustrate the stage that sediment management, and perhaps river basin management generally, is at. At this stage in the development of river basin management, of which sediment forms an important part, recognition of the need to integrate all three elements of sustainability into the decision-making process is a step in the right direction. The following sections consider how we may move forward in terms of an adaptive management framework (Section 4) and present illustrative case studies (Section 5).

4. An adaptive framework for river basin management

On the basis of the above, and information given in the other books in this SedNet-Elsevier series [1–3], it is recommended that sediment management should be part of broader land–river management, and should adopt most, if not all, of the stages described in the following sections. In addition, some key requirements are also described that, while not necessarily stages in the management process, are important nevertheless. Although these requirements and recommendations are listed below, it is important to recognize that they form part of a more circular, adaptive approach to management, which is illustrated in Figure 5. In such as adaptive framework, there are fewer rules regarding the order in which the stages are followed, and it may be that not all stages are required. The order in which the various stages occur may depend on the policy objectives and the specific basin, although in many cases the order listed below may be suitable. In other cases, it may be desirable to end the process at other stages in the cycle. Equally, it may not always be necessary to undertake all of the stages. Thus, in summary, within an adaptive management framework:

- it is possible to enter at any stage in the cycle;
- the stages do not need to be followed in a set order;

- it may not be necessary to undertake all stages; and
- it may be possible to leave at any stage in the cycle provided a suitable decision has been reached, which can include that no intervention is needed, or that further information is required. Then the process starts again.

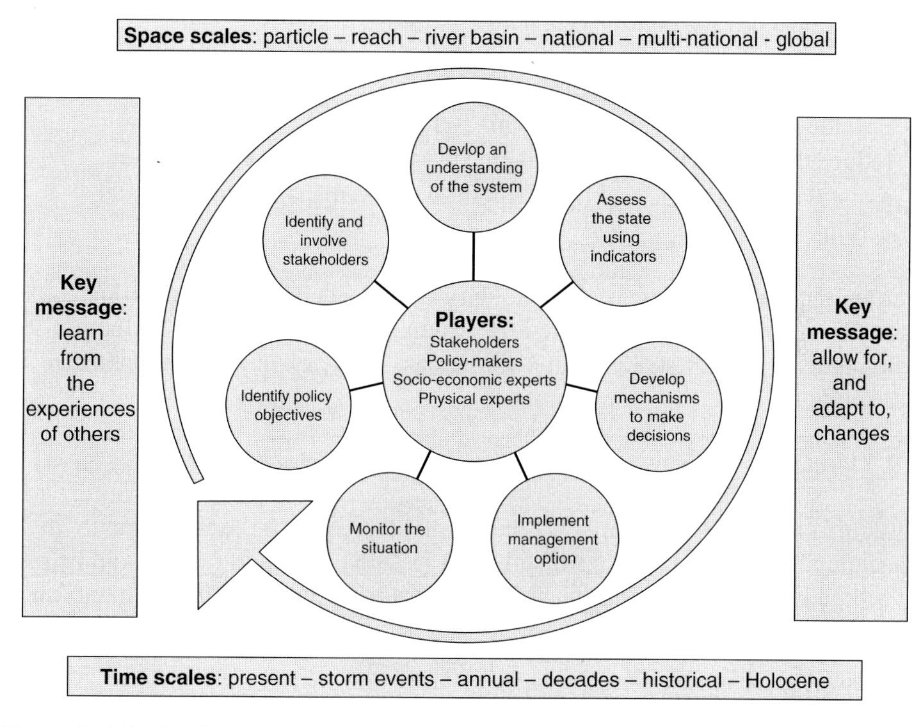

Figure 5. Adaptive framework showing the stages (outside circles) and players (inner circle) required for sediment management at the river basin scale. It may be possible to enter and leave at any stage, and there is no set order to the stages. The management framework does require an appreciation and assessment of the space and time scales involved with both the various stages and players that are relevant to the particular policy objectives. See text for further explanation

4.1. Identify the policy objectives for the river basin

Policy objectives may be regional (e.g. Rijnland water board, see Section 5.3), national (Catchment Sensitive Farming in the UK, see Chapter 3, this book) or international (e.g. Water Framework Directive, WFD; Danube basin, see Section 5.1) in nature. Importantly, it is necessary to identify and evaluate all relevant policy fields in order to ensure that objectives to meet one policy field do not compromise another policy field. Section 2 has described how policy objectives are shifting towards sustainable environmental requirements.

4.2. Involve stakeholders to help develop and implement management objectives, plans and actions

As identified in many of the chapters in this book, involving stakeholders throughout the management process is likely to be important for achieving the policy objectives. Furthermore, stakeholders should be used to:

- help formulate the policy objectives;
- provide useful perspectives, information and data;
- reduce the obstructive nature of some pressure or lobby groups; and
- assist with the long-term implementation of the management plan and actions.

Further details on the role of stakeholders are provided in Chapter 7 (this book), plus the reader is also directed to Ellen *et al.* [29] for information on communication with stakeholders. Some case study examples are presented in Section 5.

4.3. Develop a common system understanding of the river basin

System understanding may be achieved through the development of a conceptual river basin model for sediment [17, 18], which in turn should be part of a broader, integrated river basin model that incorporates other elements such as:

- surface water (quantity and quality);
- groundwater (quantity and quality);
- soil (functions; types);
- land use (including agriculture, urban, industrial, recreational); and
- (possibly) atmospheric influences and interactions.

The AquaTerra research project is an example of a research project to develop system understanding (see Section 5.2). Such an understanding may already exist for a basin, depending on the available information. If information is not currently available to permit an understanding of the system to a level such that informed decisions can be made, it will be necessary to obtain that understanding. This can be achieved through assembling data and information, which could involve one or more of measurement, monitoring and modelling (Chapters 4 and 5, this book). Stakeholders should already be involved at this stage to allow for a common understanding of the system. Typically, such information may include:

- sediment–contaminant sources;
- sediment–contaminant transport and fluxes;
- sediment–contaminant storage;
- sediment–water interactions;
- sediment–contaminant interactions;
- sediment–contaminant interactions with ecology; and
- assessment of stakeholder perceptions.

Figure 6 shows, as an example, a geomorphologically based assessment and management framework for historically contaminated river systems in England and Wales (see also Chapter 2, this book). It illustrates steps to assemble information on sediment–contaminant sources, dynamics and storage within river basins for effective and targeted management. As such, the framework shown in Figure 6 represents a very useful part of the overall decision-making process.

4.4. Use indicators to assess the state of the system

Indicators are statistical data that can be selected and observed to gain insight into the functioning of a complex system [31], and they provide a means to assess the state of a system relative to the policy objectives. Indicators help to identify where a particular system (e.g. river basin) or part of a system (e.g. river reach, lake, estuary) is within a temporal and/or spatial context. Thus, for example, an indicator may provide a means to assess how polluted a river reach is compared to that same reach under pristine conditions. Similarly, an indicator can provide a mechanism to determine if management is causing a beneficial or detrimental change in the state of the system. One indicator may be sediment quality (see [1] for further details on sediment quality assessment). Other types of indicators relevant to sediment in river basins include:

- water quantity and quality;
- aquatic biodiversity;
- aquatic productivity and/or trophic status;
- human health;
- degree of multi-functional use of a river basin;
- economic indicators (such as income); and
- social indicators (such as human happiness).

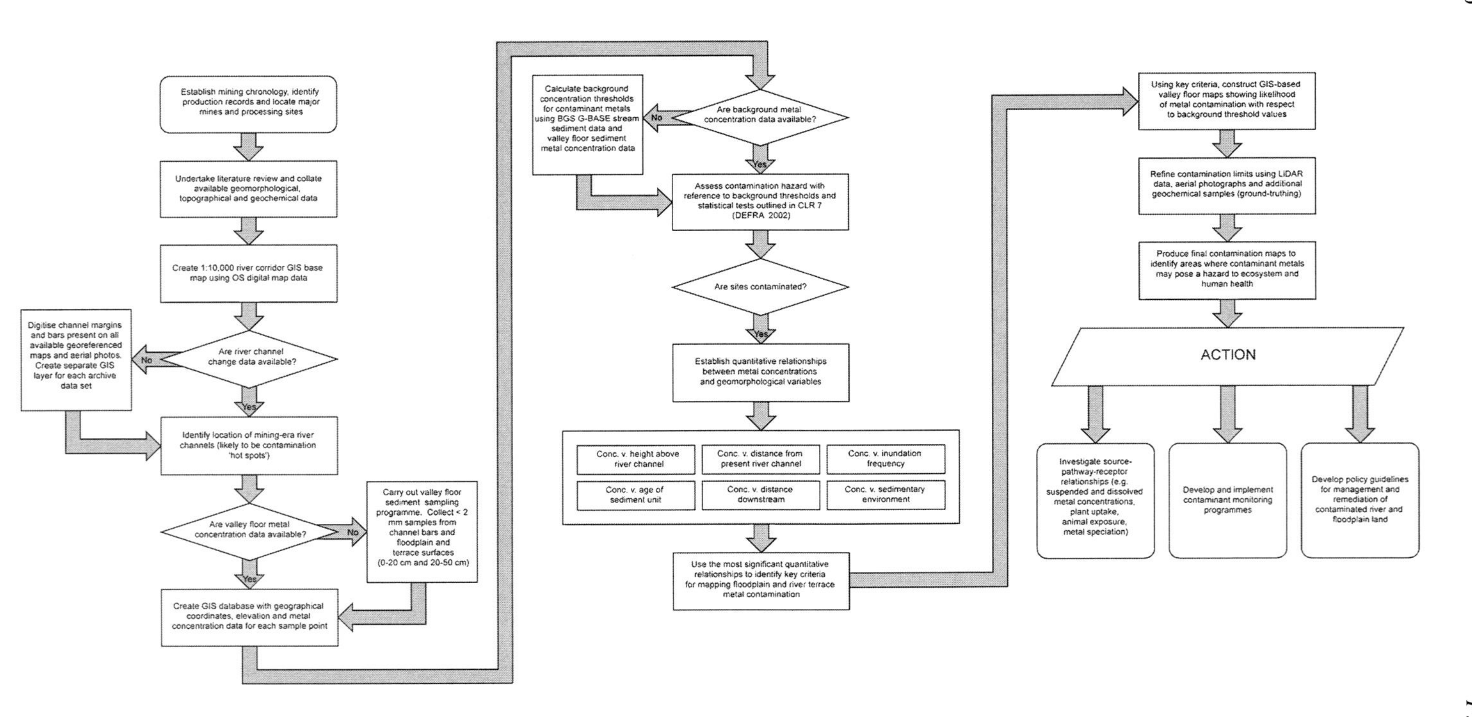

Figure 6. Geomorphologically based assessment and management scheme for historically contaminated river systems in England and Wales (source: [30], reproduced with the permission of Elsevier)

Thus, as an example, a specific species of fish that is sensitive to environmental conditions is a good indicator of the state of the system relative to the policy objectives.

A requirement for the effective use of indicators is the development of standardized and reproducible ways of measuring an indicator and of assessing the significance of the results. It is important that the insight gained into the behaviour and functioning of a system at a particular point in time or space, as provided by the indicator, can be compared to that obtained for another point in time and space. Thus in the case of the River Danube basin (Section 5.1), it is important that the indicator, and the way in which it is used, is the same in Austria as it is in Romania, otherwise the usefulness of the indicator is compromised. However, it is realized that the achievement of a commonly accepted and used set of indicators, and specifically the agreement on a 'yard-stick' for these indicators, may be one of the most challenging issues to overcome in trans-boundary river basin management.

4.5. Develop mechanisms to make decisions

At the end of the day, management is about making decisions, usually about the best option to take in order to achieve the policy objectives. In most situations, there will be more than one option (or Programme of Measures in WFD terminology), and hopefully more than one option that is acceptable. As such, there needs to be a clear and open way to reach a decision that is defendable to all stakeholders: although, of course, this does not mean that the decision is the preferred choice of all stakeholders. There are several tools or approaches that are available to help make decisions. One is cost-benefit analysis, a variation of which that is particularly relevant for environmental resources (e.g. water, soil, sediment) issues is societal cost-benefit analysis (see Chapter 6, this book) because it provides a mechanism to evaluate the concerns of stakeholders. Other types of tools relevant to resource management are risk assessment and risk analysis, and these are covered in the book by Heise [3].

4.6. Implement the management option

Once a decision has been made as to which management action to take, this should be implemented in a timeframe that is appropriate for that particular action. For example, in the case of a site-specific action, such as dredging of a reservoir or harbour, then this could be done reasonably quickly and at a rate determined by the deposition of sediment, with the action requiring days to months to implement. In the case of the decision to control widespread soil erosion and sediment delivery from agricultural land throughout a river basin, the selection of this option is just the start of a long and complex process that may require years or decades to implement, especially if new legislation or financial incentives are required.

It could be that the management decision was that no action should be taken. This is sometimes referred to as the 'zero option' or 'zero alternative'. In many cases, the decision to take no action is suitable because there may not be the required understanding of the system. In other words, if those involved with making the decision are not certain that the list of available options will result in a better situation, then it may be best not to intervene at that moment. It may also be that no action is required because the system will naturally correct itself. A pulse of sediment into a river channel from the collapse of a section of channel bank may result in a site-specific problem, but most rivers are able to transport any deposited sediment away naturally during high flow conditions and the overall downstream impact on the river may be insignificant. Indeed, such events are a natural part of the normal geomorphological functioning of river basins, and are required for a variety of reasons, such as helping to maintain certain aquatic habitats. Hence, it is important to assess the need to manage such natural processes in light of the benefits they offer.

4.7. Monitor the situation

Monitoring of the situation represents an important part of the overall decision process and should feature at several levels. Monitoring is likely to be part of the early stages of the decision-making process when common system understanding is required (Section 4.3). Thus monitoring may be required to assemble relevant information on sediment behaviour and functioning, such as to provide data on sediment fluxes and sediment quality (see [19, 20], Chapters 4 and 5, this book). Monitoring is also something that is likely to be required in the latter stages of the decision-making process, after an initial management decision has been made. In this situation, monitoring of sediment behaviour and function is required to assess that the management decision was correct. It may be that no further action is required, or that the management option was not suitable, in which case it may be necessary to repeat part of the decision-making process (Figure 5). This use of post-action monitoring to provide a feedback mechanism to evaluate the response of the system to a management action is nicely illustrated in Figure 7.

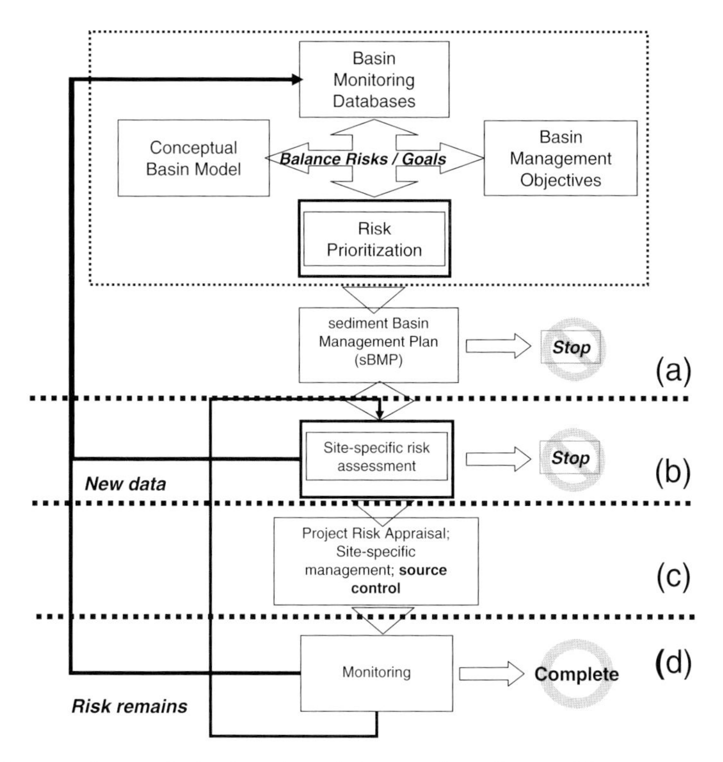

Figure 7. Process diagram for basin-scale and site-specific sediment risk management. Note the use of monitoring at the initial stages to inform prioritization and decisions, and during the latter stages of the process to assess the outcome of the action and the associated feedback mechanism for further assessment and action if required (source: [32], reproduced with the permission of Elsevier).

4.8. Allow for change

Throughout all of the stages above, a key element is to allow for change. If anything can be certain, it is that there will be some form of change within the system, i.e. the river basin. There are several forms of change – environmental, economic and social – and there are several space and time scales over which these changes occur. At the moment, there is great concern over global and regional environmental changes, of which climate change is perhaps the most prominent. The effects that such changes will have on soil, water and sediment resources in river basins is likely to be profound, and will influence how we manage these resources and how we adapt to these changes. This allows us to understand the system better.

4.9. Learn from the experiences of others

Learning *from* others is a very useful part of the overall process that is often neglected. Indeed, one of the best ways of assessing the likely success of a particular management approach or action is to look at cases where this, or something similar, has been used, and to draw lessons from its success or failure. Thus, learning from the experiences of others should represent an important stage in the overall decision-making process. On the other hand, learning *with* others – stakeholders, scientists, policy-makers and decision-makers – is an important prerequisite for the process to understand the system. It also serves to inform about the interests and ways of thinking of the other actors involved. This social learning process builds the 'social tissue' that is needed to generate common system understanding and to come to decisions that can be supported by the actors involved.

The focus of this book, and indeed the SedNet-Elsevier book series [1–3], has been sustainable sediment management in Europe. It can be argued that many exciting developments are happening in Europe, driven by EU and national policy and research initiatives, and this book describes many of these, such as:

- the Water Framework Directive;
- forthcoming EU soil and marine policies;
- SedNet;
- AquaTerra (see Section 5.2); and
- the work of the International Commission for the Protection of the Danube River and UNESCO-ISI in the River Danube basin (see Section 5.1).

However, there is also much to be learnt from the experiences of other countries. The USA, for example, has long been a pioneer in water management at the scale of the river basin or 'watershed', and has been undertaking sediment research and sediment management as part of this. Some of the main sediment programmes and initiatives within the USA include SuperFund, the National Sediment Quality Survey and the work of the USEPA, NOAA and the Army Corps of Engineers. Many of these have been successful, particularly in terms of contaminated sediment management, while others have been less so, often due to the fragmented way in which sediment has been addressed within river basin management. In Australia and New Zealand, there are also many useful lessons to be learnt. Morgan [33], for example, describes the positive effects that the Land Care movement in Australia has had on controlling soil erosion and sediment delivery. Land Care groups have the following characteristics [33]:

- they tackle a broad range of issues, providing an integrated approach to resource management;
- they are based on neighbourhoods, usually covering catchments with contiguous boundaries, rather than just groups of landowners with a common interest;
- their impetus comes from the community, providing 'ownership' of any programmes for erosion or sediment control and improving coordination, collaboration and communication between various stakeholders; and
- they can produce and implement proposals which are economically realistic within the funding support mechanisms available.

Similarly in Canada and other countries, environmental stewardship programmes have shown promise for integrated environmental management at basin and landscape scales by providing a mechanism for stakeholder involvement throughout the decision-making process, and in providing a sense of ownership to a particular environmental issue. Such an inclusive approach is particularly important in countries like Australia, Canada and New Zealand, where First Nations or aboriginal people can have important land use claims and have a wealth of useful information on land and river management that can benefit the decision-making process.

5. Case studies linking environmental policy developments, social and physical sciences, and river basin management

The following sections provide four case study examples that illustrate the exciting ways social and physical sciences are playing a major role in developing advice and plans for river basin management, of which sediment forms an important part. They illustrate some of the steps in the adaptive management framework identified above. The first case study (Section 5.1) describes a major initiative within the River Danube basin, which is largely driven by the need to implement the WFD and other environmental policy fields. It illustrates the requirements (and difficulties) for system understanding within a large and complex multi-national river basin. Section 5.2 describes how physical and social scientists are interacting within a major, current EU research project, AquaTerra, which is also linked to the activities within the Danube basin. The Rijnland case study (Section 5.3) provides a good example of the role of stakeholder processes in relation to knowledge utilization and water and sediment management in the Netherlands. Section 5.4 describes an environmental stewardship programme, the Fraser Basin Council in Canada, which provides an example of the lessons that can be learnt from the approaches and experiences of others.

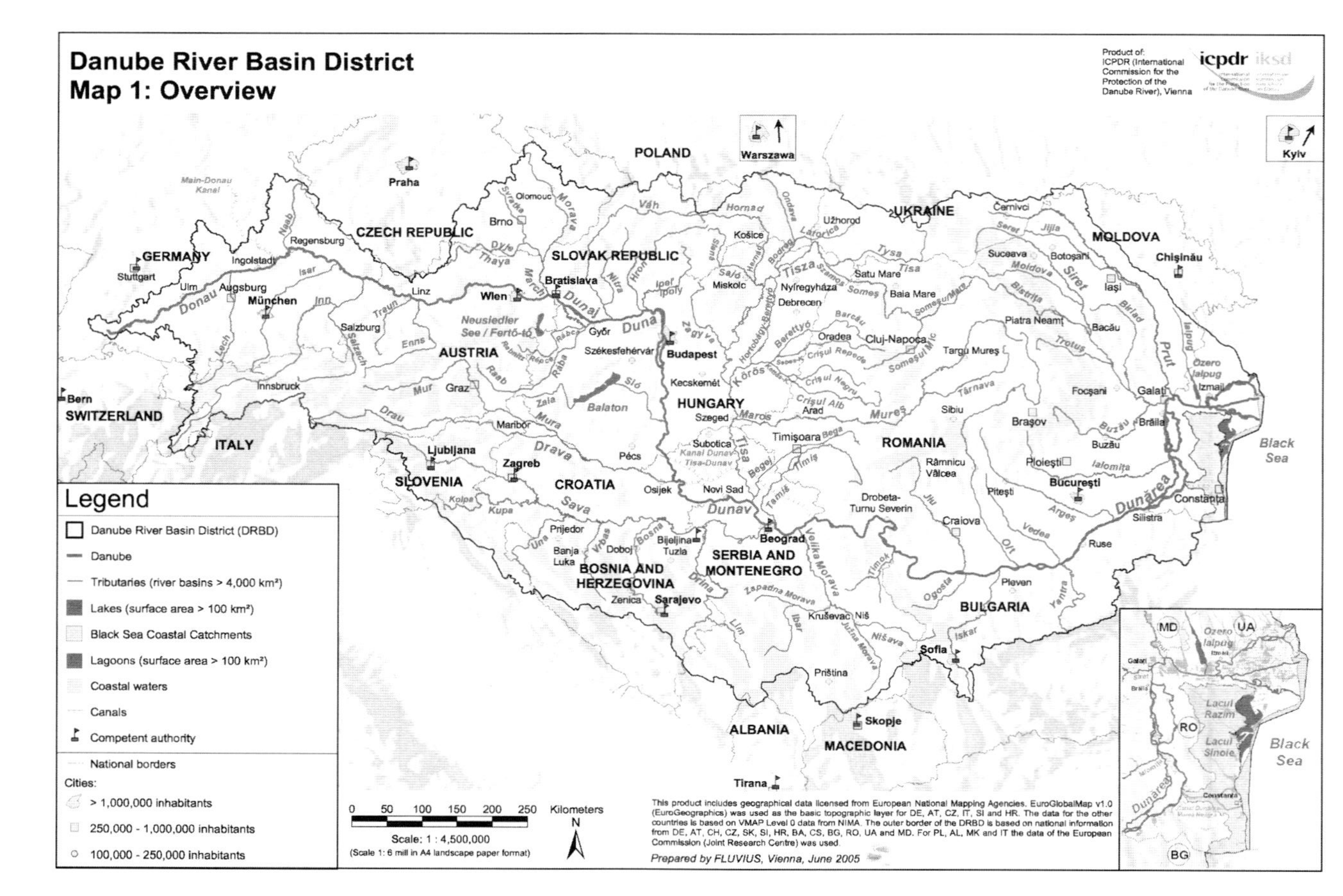

Figure 8. Map of the River Danube basin (source: [36])

5.1. Case study 1: Sediment initiatives within the River Danube basin

The Danube basin is the second largest river basin in Europe, covering 801,463 km^2, and presently consists of 19 countries (including EU-Member States, Accession Countries and other states that have not applied to join the EU), making it the most international drainage basin in the world [34]. It lies to the west of the Black Sea in central and south-eastern Europe (see Figure 8). The Danube basin has a tremendous diversity of bioclimatic habitats through which rivers and streams flow, including glaciated high-gradient mountains, forested middle mountains and hills, upland plateaus and plains, and lowland wetlands near sea level. In addition, the basin has a population of over 81 million people and is heavily industrialized in places. In recognition of the difficulty in managing water within the Danube basin, the International Commission for the Protection of the Danube River (ICPDR) was established in 1999 [34]. Subsequently, the UNESCO International Sediment Initiative (UNESCO-ISI) has become actively involved with the ICPDR on addressing sediment issues in the Danube basin.

Figure 9 shows that the Danube basin is at risk of failing to achieve the environmental objectives of the WFD [35], due to pressures from hydromorphological alterations, nutrients, hazardous substances and organic pollutants [36]. Although it is likely that each of these significant water management issues is also linked to sediment, based on the available information to date it has not yet been possible to estimate whether there is also a risk of failure to meet the WFD objectives due to sediment problems. Therefore, the ICPDR river basin managers decided that any consideration of sediment management as a significant water management issue has to be preceded by further investigation and the collection of data in order to provide an improved understanding of the sediment system in the Danube at the river basin scale. A thorough assessment of sediment quantity and quality issues within the Danube basin should be available by the end of 2008 as a document accompanying the WFD Programme of Measures. The aim of this exercise is to determine whether sediments should become a significant water management issue in the second WFD implementation cycle. The Danube basin, and particularly the work of the ICPDR and UNESCO-ISI, therefore, provides a good example of the steps currently being undertaken to assess the role and function of sediment within broader river basin management for EU environmental policy implementation, within perhaps the most complex river basin in Europe.

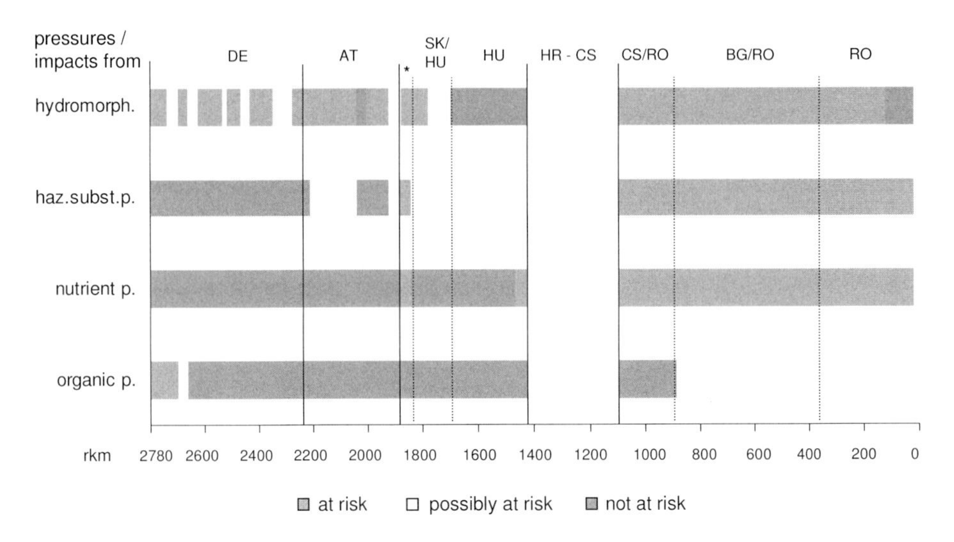

Figure 9. Causes of risk of failure to meet the objectives of the EU Water Framework Directive in the Danube basin. Pressure/impacts are from hydromorphology, and pollution from hazardous substances, nutrients and organics. The countries on the top axis are Germany (DE), Austria (AT), Slovak Republic (SK), Hungary (HU), Croatia (HR), Serbia (CS), Romania (RO) and Bulgaria (BG). The darker shading represents "not at risk" (source: [36]).

At present, there are several key management issues, covering both sediment quantity and quality aspects, for consideration in the Danube basin, and these are described below. For further information, see the reports of the ICPDR [36, 37] and SedNet [38].

5.1.1. Sediment quantity and quality issues in the Danube basin

Sediment transport

According to the Danube Analysis Report [36], regulation works during the 19th century, together with the nearly complete loss of sediment supply from the upper Danube basin in the 20th century (retained by a series of dams from the Alps down to Gabcikovo hydropower dam), reduced downstream sediment fluxes. In turn this has increased the sediment deficit for the entire Danube up to the Iron Gate dams (two major hydroelectric dams built at a gorge dividing the Southern Carpathian Mountains in the north and the Balkan Mountains in the south and separating the middle and lower Danube basin) and beyond. However, recent data on suspended sediment transport in Austria [37] show that the present values are as high as they were in the early 1960s, and this is supported by data on suspended sediment concentrations in the upper Danube from the ICPDR Transnational Monitoring Network. These contradictory facts on sediment fluxes in the upper Danube show a clear need for further investigation of sediment dynamics in order to achieve a common consensus of the situation. In the lower Danube, a reduction in the discharge of suspended sediment to the Black Sea was recorded at the Isaccea section, which lies upstream of the Danube delta, where it has decreased by about 50% during the last 100 years.

A serious problem in an assessment of the sediment flux in the Danube basin is the relatively high uncertainty associated with such quantification. For example, during an extreme flood event in August 2005 on the River Inn at Innsbruck, Austria, the sediment flux observed within five days was more than twice as high as the annual sediment load reported for the same site for the whole of 2004.

Influence of dams – sedimentation and channel incision

Upstream of a dam sediment is retained and often has to be extracted to maintain river depth for hydropower generation and navigation. For example, gravel extraction of approximately 15,000 m^3 $year^{-1}$ is necessary on the River Traun on the impounded section of Abwinden–Asten in Austria, which acts as a sediment trap. Downstream of dams the loss of sediment transport requires artificial additions of sediment to stabilize the river bed and to prevent incision. This is the situation downstream of the Freudenau dam, where the addition of 160,000 m^3 $year^{-1}$ of bed load is required. Immediately downstream of the Iron Gate dams, incision of the riverbed is monitored as a result of changes in flow and sediment regime. The overall reduction of sediment transported by the Danube over the long term leads to intensive erosion on unregulated banks and islands in the lower Danube region, e.g. Tcibtriza Island, Belene Island, Garla Mare, Calafatul Mic and Cama-Dinu.

The interruption of the longitudinal river continuum due to dams, and the retention of sediment in the impounded stretches, also impacts the riverine ecology. As a consequence of reduced slope and current velocity, fine sediments cover the natural habitats of the bottom-dwelling organisms and clog the interstices in the bed sediments. This leads to a diminished flow of oxygen into the bed sediments and to a reduced recharge to, or inflow from, groundwater. These changes in flow and substrate composition affect the benthic invertebrates and the spawning grounds for fish (also see Chapter 1, this book).

Dredging

One of the main pressures resulting from navigation is related to channel maintenance. Studies have shown that on the Austrian Danube up to 60% of the channel incision in several sections downstream of Vienna was caused by increased regulation and dredging activities for securing waterway transport. However, a recent ruling by the Austrian Supreme Water Authority only permits dredging in the Danube if no more than 50% of the dredged material is used for structural measures on the riverbanks. The rest of the material is deposited in the river such that it can be continuously mobilized by the flow.

In the lower Danube region, lateral riverbed erosion dislocates the navigation channel in the Danube. Additional river training works, as well as dredging of shallow fords to maintain the minimum shipping depth, are carried out. In the Danube delta, dredging is also an important problem. At the start of the 20th century, but especially in the years 1960–1990, many canals were dredged in the area of the delta to optimize water circulation needed for an increase in fish farming.

Sediment quality

There is only a limited amount of information on the contamination of sediments by priority substances in the Danube basin; the lack of data is especially unsatisfactory in the lower Danube. The characterization of sediment quality in the Danube is primarily based on the results of special surveys, such as the Joint Danube Survey organized by the ICPDR, and scientific projects. In terms of nutrients, especially for the phosphorus (P) balance, sediment plays an important role. The Iron Gate backwater area represents a major in-stream storage area due to the net deposition of P attached to sediment particles. Recent research indicates that about one-third of the incoming load is stored semi-permanently [36].

5.1.2. Future steps in sediment management in the Danube basin

With the view of obtaining new information and understanding of the role, quality and fluxes of sediment in the Danube basin, in 2006 the ICPDR agreed on a series of operational requirements concerning both sediment quantity and quality [37]:

Sediment quantity

- To clarify the sediment deficit problem in the basin. As described in the Roof Report 2004 [36], it is necessary to investigate further the sediment fluxes in the basin from a long-term perspective, and to establish a sediment budget for the basin.
- Special attention should be given to the role of floods in sediment transport, since a substantial part of an annual sediment load can be transported by flood events.
- An appropriate assessment of the sediment balance necessitates the collection and analysis of long-term data sets. An international consensus has to be achieved on this issue.

- The intercomparability of the existing data sets on sediment quantity has to be investigated and quality assurance criteria have to be agreed. The sediment sampling methods in use in the Danube countries have to be compared and assessed with the view of achieving a common standard.
- The environmental aspects of dredging have to be discussed at the basin scale level. This process should lead to the adoption of recommendations.

Sediment quality

- The existing data on sediment quality in the Danube basin – of primary concern is the occurrence of priority and other substances as defined by the WFD – should be collected and reviewed. This process should apply to channel bottom sediment as well as to suspended solids.
- Attention should be given to the availability of metadata focusing on sampling techniques, analytical methods and analytical quality control approaches, including the assessment of errors and uncertainty.
- Based on an analysis of existing data, further information on sediment quality should be collected. This refers primarily to data on concentrations of priority and other substances in the solid phase.
- An inevitable prerequisite of assessing the level of pollution of sediments and suspended solids is the establishment of environmental quality standards as required by Article 16.7 of the WFD.
- There is a need to improve the understanding of the role of sediment in the functioning of the natural sediment–soil–water system in the Danube. The actions should be aimed at assessing the combined impact of sediment quantity and quality on the ecological status.

Key sediment research actions

In line with the above mentioned operational requirements, several key actions were proposed in order to improve the background knowledge on sediment quantity and quality issues in the Danube basin:

- Sediment balance. UNESCO included in its priorities [39] to support the ICPDR in the determination of the Danube sediment balance at the river basin scale.
- Dredging issues. Preparation and adoption of recommendations on the environmental aspects of dredging in the River Danube basin are planned. This process should be done in cooperation with the Permanent Environmental Steering Committee of the Central Dredging Association (CEDA).
- Collection and evaluation of data on sediment quality. The process of collection of existing data on the quality of sediments and suspended solids

in the Danube basin, and of acquisition of the missing information as well as of the necessary metadata, will be managed by the ICPDR. New data will be obtained via planned international activities, such as the United Nations Development Programme/Global Environment Facility (UNDP/GEF), Iron Gate project or the Joint Danube Survey 2. However, it is recommended that the results of national monitoring programmes should also be compiled.

- System understanding. This process aims at achieving a better understanding of the role of sediment in the functioning of the natural sediment–soil–water system in the Danube in light of the WFD environmental objectives. The knowledge should be increased by project-oriented activities. The major topics to be dealt with are shown in Figure 10. It is recognized that existing activities such as SedNet (see [38]) and AquaTerra (the Danube is one of the case study basins, see Figure 11) could be very helpful through the collation and synthesis of existing experience and knowledge on the impact of sediment quantity and quality on ecological status.

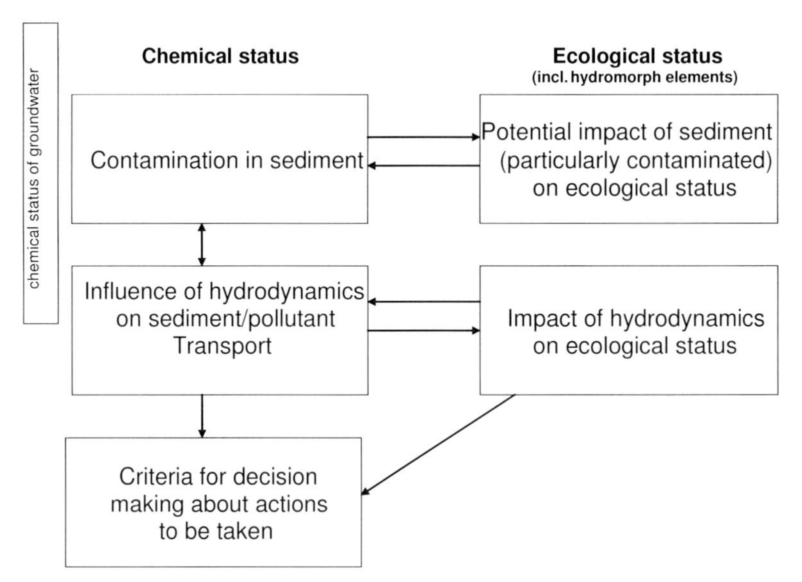

Figure 10. Structured goals of proposed activities towards a better understanding of the natural sediment–soil–water system within the River Danube basin (source: [36, 37])

5.2. Case study 2: The EU AquaTerra research project

The AquaTerra project [40] is a multi-disciplinary project in the EU 6th Framework Research Programme (FP6) involving 42 partner organizations from 12 EU and several other European countries, that runs from June 2004 to May 2009. AquaTerra aims to 'provide the scientific basis for improved river basin management through better understanding of the river-sediment-soil-groundwater system as a whole, by integrating both natural and socio-economic aspects at different temporal and spatial scales' [40] (Figure 11). As such, understanding sediment behaviour and dynamics, particularly in terms of sediment–water and sediment–chemical interactions, forms a key part of the AquaTerra project. This project is being driven by current and forthcoming environmental policy needs by developing state-of-the-art process understanding, which in turn will help to implement current policy and also help to formulate future policy development for river basin management. As such, AquaTerra represents one of the most recent, largest and ambitious research projects in Europe trying to link scientific understanding to environmental policy.

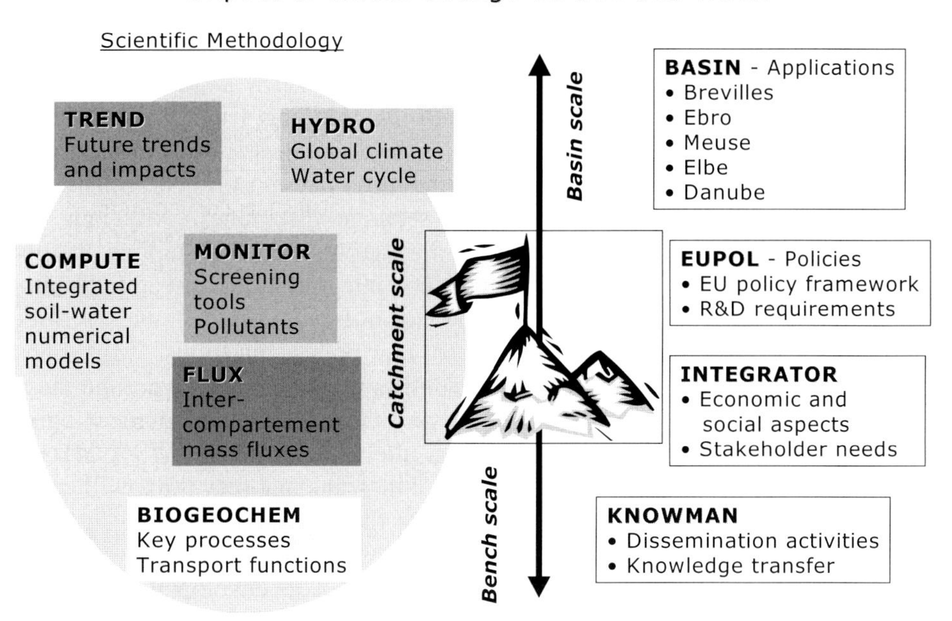

Figure 11. The EU project AquaTerra (source: [40])

Regarding the language problems that originate from the many different disciplines and countries involved, one of the main challenges in AquaTerra is to come up with meaningful results for policy. Therefore, one of the work packages of AquaTerra, EUPOL, aims to find a way to link policy demands to the scientific information on processes being generated by AquaTerra for river basin management. To show the linkages between the scientific knowledge of the river basin system generated by AquaTerra to both the current and likely future needs of policy-makers and stakeholders, a scientific framework was developed. The framework is a matrix that links individual research deliverables of the AquaTerra subprojects to the needs of stakeholders and policy-makers, and vice versa. In order to populate the scientific framework, the EUPOL research team sent out a questionnaire to the leaders of each of the AquaTerra work packages to collect more detailed information about the deliverables as currently envisaged in the context of policy benefits. Secondly, the EUPOL team organized brainstorming sessions and interviews with policy-makers in different countries. The brainstorming sessions resulted in a list of questions that could be raised by policy-makers now or in the future. In total, the brainstorming sessions yielded 54 policy questions that were most commonly related to land use, river basin management, water and climate change, while questions related to agriculture and soil were the least common. According to the scientific framework, a relevant amount of policy questions could be answered very well by some of the work packages of AquaTerra [41].

As stated earlier, one of the challenges in AquaTerra is to integrate the research, and the results from the research, by bringing scientists from very different disciplines together to generate a better understanding of the river basin system. The AquaTerra research community, like most others, consists of many smaller factions, grouped around separate disciplines. Although the AquaTerra researchers share a scientific background, each discipline has its own practice and language, which hampers integration of research results. As the EUPOL team knows from the scientific framework which work packages can significantly contribute to the policy questions, workshops between policy-makers and AquaTerra scientists will be organized around these questions to improve the integration of results and to generate meaningful results for policy-makers. To improve the integration further, the AquaTerra community is asked to think about the central message that they think will come out of AquaTerra and what kind of evidence they have for this message and how this relates to policy-making. A series of workshops with the AquaTerra research (and policy) community will be organized to develop this message during the last two years of the project. Significant efforts are, therefore, being made to connect to policy-makers and to integrate the results of this big project. When the projects ends in 2009, it will be known whether these attempts at integration have been successful.

5.3. Case study 3: The Rijnland dredging project, The Netherlands

Water boards are the oldest governmental bodies in the Netherlands, founded in the 13th century, to govern the amount of water in a specific area for all land owners, with the aim of preventing conflicts over water issues. Rijnland governs the polder areas in the west of the Netherlands close to the sea, roughly between the cities of Gouda, the Hague and Haarlem. As water boards are responsible for the water quality and quantity issues in their areas, they are also responsible for dredging the polder canals and ditches. The Rijnland water board has not dredged sediment from the water network in the polders for decades, resulting in a backlog, which in turn has led to calls for a major dredging operation. The accumulation of sediment compromises both the quantitative and qualitative aspects of water management, diminishes ecological life, and hinders economic and recreational use of the water bodies. To avoid public resistance at the onset of the dredging activities, the Rijnland water board wanted to establish stakeholder involvement. Rijnland and TNO agreed to apply a newly developed methodology for stakeholder involvement designed specifically for sediment management.

Starting from the knowledge that there existed different perspectives on sediment management (see [29]), TNO developed a stakeholder involvement process for sediment management, based on experiences with the development of the European Awareness Scenario Workshop Methodology (http://cordis.europa.eu/easw/home.html) and with interactive decision-making processes concerning CO_2 reduction in housing projects [42]. The process framework consists of seven steps. The first step is the convening step, i.e. getting people to the table. This entails: assessing the situation (a description of both the natural and the social systems); identifying and inviting the stakeholders; locating the necessary resources; and organizing and planning of the process [43]. The first step also includes the design of the communication process. The second step is problem analysis and the composition of process rules, in order to clarify the goal and expectations concerning the process and the role of the stakeholders. This involved a workshop organized to get an overview of specific issues, causes and effects related to sediment that are important for the stakeholders. The third step focuses on dreams and nightmares for the future. With the problem analysis of the second step in mind, the goal of the third step is to develop the images of the future of sediment management. The fourth step is aimed at asking questions and exchanging knowledge. In this workshop the stakeholders get the opportunity to ask questions of and enter into a dialogue with experts from science and practice in a 'knowledge market'. The fifth step focuses on possible options for sediment management. The goal of this step is to get an overview of which sediment management options can be used to fulfil the desires of the different perspectives. The sixth step is used to find mutual benefits, by discussion in subgroups, which are focused on similarities rather than differences. The result of this workshop is an overview of sediment management options, including the conditions in which these options can, or cannot, be used, which will be handed over to the responsible decision-maker at the end of the workshop. In the final, seventh step, the decision-maker will present the outcome of their decision process: which options they have chosen; what conditions are applied; and, most importantly, the rationale as to why these choices were made. The stakeholders get the opportunity to discuss these choices. With this seventh step the process is ended, but this is only the start of the actual work. It is essential to keep the stakeholders involved in the project, for example by inviting them to visit the actual location of concern.

At the beginning of 2006, representatives from the different stakeholder groups, such as farmers, resident committees, environmentalists, recreation associations, angling associations, local authorities etc., were invited to discuss how the dredging operation should be executed and how to dispose of the dredged material. The process finished at the end of 2006. The first two workshops were executed according to the plan, but the water board felt that the process was too slow. In the next workshop they wanted to present some technical solutions that were available. So during the third workshop, which was aimed at asking questions and exchanging knowledge, the stakeholders, who thought that they had been invited to a 'knowledge market', were asked to give their opinions about the technical solutions that were presented by Rijnland. This approach caused a small uproar. The stakeholders felt manipulated by Rijnland and it appeared to them that Rijnland had already decided what to do. As soon as the protests were expressed, a 'time-out' was called and the problems were discussed. It was agreed that a small group of representatives from the stakeholder group would discuss the process separately with Rijnland. In the separate meeting that was held afterwards with a delegation of the stakeholders, they expressed that they wanted to respond to location-specific solutions. They suggested that the water board should present these solutions to local people who have local knowledge of the areas. Furthermore, the stakeholders wanted to separate this discussion from the long-term discussion on sediment management. These suggestions for redesign of the process were discussed in the following workshop and were implemented after approval by all stakeholders and the water board. With the help of the local municipalities, the water board listed several options for disposal of the dredged material. To find out whether any options were missed, and under what conditions these options were acceptable for the stakeholders in the area, two local workshops were organized that were attended by local people. The process ended with a clear, appreciated and approved list of options and conditions, providing a sound foundation for the disposal of the dredged material. The list contained more disposal options than were originally identified by Rijnland, so the water board were very satisfied with the list. During a subsequent meeting, the list of options was officially handed over to the administrator of the water board in December 2006.

The Rijnland case is an example of stakeholder involvement in which the input of local knowledge, combined with scientific knowledge, was crucial. By inviting local people to the workshops, people who have relevant local knowledge of the area (farmers, environmentalists, etc.) were able to generate new options for the dredging activities and disposal of the dredged material.

5.4. Case study 4: The Fraser Basin Council environmental stewardship programme, Canada

The Fraser Basin Council (FBC) in Canada is a useful example of an environmental stewardship programme. The FBC is a non-governmental, not-for-profit organization that was established in 1997 to provide a mechanism to contribute to the sustainable management of the Fraser River basin [31]. The Fraser basin is 220,000 km^2 in area (i.e. about the size of Romania or the UK) and has a population of ca. 2.8 million people, most of whom are located in the city of Vancouver and surrounding area. Although the basin is fully contained within the province of British Columbia, thereby removing cross-boundary and multi-national issues, there are many potentially conflicting concerns for how the basin is managed – particularly with regard to land and river management effects on water quantity and quality. At the core of the FBC are the involvement of all stakeholder groups and the undertaking of research and other activities. A recent initiative has been to develop a series of 40 'sustainability indicators' that provide a measure of the state of the basin according to social, economic and environmental policy objectives. Key environmental indicators include a water quality index, and freshwater (e.g. wild salmon) fish stocks and habitats. While there are no indicators specific for sediment, sediment quantity and quality is recognized through its influence on water quality and ecological quality.

Furthermore, the FBC has also been pro-active in promoting, facilitating and funding research that it believes is relevant for the sustainable development of the Fraser basin. For example, significant amounts of gravel are deposited each year in the lower Fraser River during the spring runoff. Gravel movement and build-up in some areas of the river reduces the ability of local communities to protect themselves from floods, and also affects commercial navigation in the river. However, the removal or movement of in-channel gravel has implications for fish habitat as well as in-channel infrastructure such as bridges, pipelines and bank protection. To address this complex issue, the FBC brought together all interested parties to facilitate the development of a plan that would address key issues including flood and erosion protection, fish and aquatic habitat, navigation, First Nations' concerns and gravel resources. The resulting 5-year plan developed by the Fraser River Management Plan Steering Committee defined the location, timing and quantity for potential gravel removals in order to focus efforts on flood, erosion and navigation hazards, while avoiding impacts to habitat. The agreement includes approval to remove up to 500,000 m^3 of gravel in each of the first two years of the plan and up to 420,000 m^3 in each of the following three years. The agreement also specifies monitoring requirements in order to protect fish and aquatic habitat, and the development of a River Management Fund that puts net royalties from gravel removals back into river management. The FBC has also convened and facilitated research into hydraulic modelling of flood risk in the lower Fraser basin, including assessing the role of dredging of river bed sediment on flood risk.

The FBC is a good example of an initiative aimed at providing 'social well-being supported by a vibrant economy and sustained by a healthy environment' [31] that operates at the river basin scale. Other river basin stewardship programmes in the Fraser basin include the Rivershed Society of British Columbia [44], and relevant documents that include guidance on sediment management as part of river management include *Stream Stewardship* by the Ministry of Environment, Lands and Parks [45].

6. Conclusions and recommendations

The new era of sediment management needs an integrated and holistic approach that is characterized by a high degree of complexity. This new approach at the river basin management scale requires a new form of knowledge production that should acknowledge the multiple rationalities and different viewpoints that are brought in by the variety of stakeholders that are involved. This new knowledge production should contain multi- or transdisciplinary research methods, involvement of stakeholders in the research process, and a learning approach in river basin management.

In the cases that are presented in this chapter, some elements of this new way of producing knowledge are shown. In the AquaTerra case, it is highlighted that the integration of different policies (water, soil, groundwater, sediment) in river basin management presents a challenge to science to better integrate the research from different relevant disciplines in order to understand the river basin system as a whole. Researchers should work together in multidisciplinary teams from the beginning and take time for the development of new theoretical multidisciplinary scientific frameworks and for the process of interaction between the disciplines. In the Rijnland case, the value of the production and use of knowledge in stakeholder processes is highlighted, especially the value of local knowledge. The input of local knowledge was very valuable in the process and produced a variety of local options to dispose of all the dredged material and gave hints on how to dredge in a more sustainable manner. The importance of stakeholder engagement and of encouraging ownership of the environment and its management through environmental stewardship was further demonstrated by the Fraser Basin Council.

The research to substantiate river basin management should allow for a learning approach with short feedback loops that will progress the understanding of the river basin system from different perspectives. Monitoring of the river basin system is a key element in this learning approach, as from monitoring all involved stakeholders can learn how the river basin system behaves. Subsequently, the measures that are needed to bring the river basin towards a better state can be discussed and implemented. By continuous monitoring, the effects of these measures on the river basin system can be observed and adjustments to the measures can be proposed, if needed.

Active and inclusive stakeholder involvement, and research for understanding the river basin system, represent stages in a cyclic, adaptive management framework. Section 4 describes these and other stages and presents a conceptual representation of this framework for the river basin scale (Figure 5). Figure 12 presents another example of a cyclic framework for river basin management aimed at the linkages between system understanding and policy development and management. These frameworks (Figures 5 and 12), and the others presented in Chapter 2 (this book), are all valid and provide a mechanism to help decision-making. It is tempting to try to develop a single, all-encompassing framework that can be used to manage all sediment-related issues in all river basins. Due to the great variety of sediment issues and great variability in river basins it is likely that such a goal may be elusive. Nevertheless, the frameworks presented in this book, including Figure 5, should help to conceptualize sediment behaviour and dynamics in river basins, and help the formulation of management plans.

There are clearly areas where further research is required in order to provide the understanding for sustainable sediment management as part of the broader management of environmental resources. The list below, while not exhaustive, identifies some of these research and understanding requirements, and is partly based on recent recommendations by SedNet [38, 47]:

- improving our understanding of the relation between sediment contamination and its actual impact on the functioning of ecosystems (ecological status);
- improving our understanding of the combined impact of sediment quantity and quality on ecological status;
- improving our understanding of the roles of local and global environmental change on system response;
- developing ways to differentiate between the effect of measures and natural variability, specifically over a long time frame;

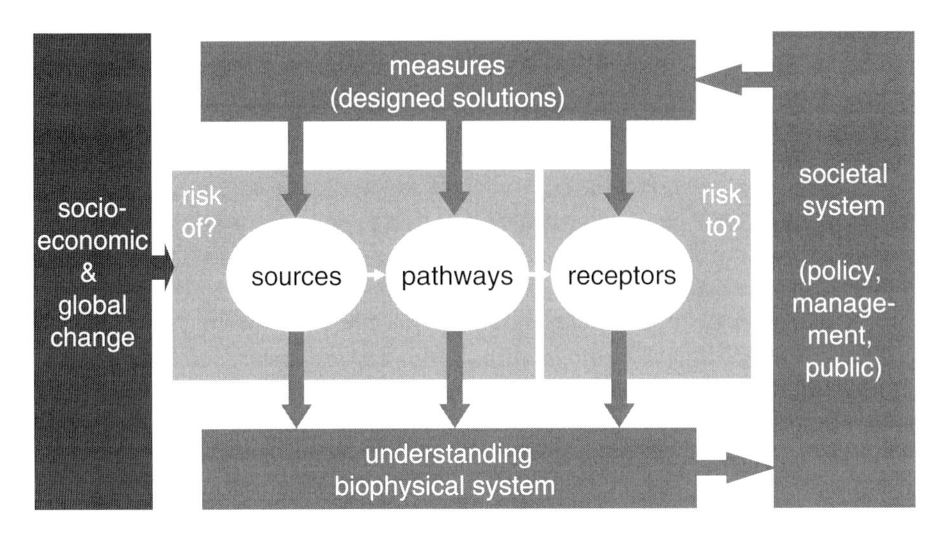

Figure 12. A conceptual framework for system understanding and its application in river basin policy development and management. Note that the framework is explicitly cyclic and hence iterative, thus driving improvements. Abbreviations: xn = spatial scale (multiple sources, pathways and receptors), Δt = temporal scale (e.g. climate change and changes in socio-economic driving forces) (source: [46])

- developing improved methods to link physical, economical and societal assessment of sediments; and
- developing improved methods to link sediment dynamics in land, riverine, lacustrine, estuarine and marine environments.

In summary, the chapters in this book and the others in the SedNet-Elsevier series [1–3] have reviewed key issues regarding the sustainable management of sediment resources in river basins. While research on, and management of, sediment have long histories, it is only relatively recently that sediment has been identified as an important component of broader land and water management. In many respects, guidance for sediment management lags behind those for other, similar resources such as water and soil. It is only through recognizing the interaction and connectivity between soil, water and sediment, and other resources such as air, and the impact that these have on ecological and human health, that we can hope to manage environmental resources in a sustainable way. It is hoped that these books contribute to this process and encourage further developments in this area.

Acknowledgement

We would like to thank Sue White for comments on an earlier version of this chapter.

References

1. Barceló D, Petrovic M (Eds) (2007). Sustainable Management of Sediment Resources: Sediment Quality and Impact Assessment of Pollutants. Elsevier, Amsterdam.
2. Bortone G, Palumbo L (Eds) (2007). Sustainable Management of Sediment Resources: Sediment and Dredged Material Treatment. Elsevier, Amsterdam.
3. Heise S (Ed.) (2007). Sustainable Management of Sediment Resources: Sediment Risk Management and Communication. Elsevier, Amsterdam.
4. Spliethoff H, van der Kolk J (1991). Environmental management: developments and interactions between technology and organisation. In: Mol, Spaargaren, Klapwijk, Technology and Environmental Policy (Technologie en Milieubeheer). SDU, Den Haag.
5. Keijzers G (2000). The evolution of Dutch environmental policy: the changing ecological arena 1970–2000, and beyond. Journal of Cleaner Production, 8: 179–200.
6. Boons F, Baas L, Bouma JJ, de Groene A, LeBlansch K (2000). The Changing Nature of Business. International Books, Utrecht, The Netherlands
7. Simons L, Slob A, Holswilder H, Tukker A (2001). The fourth generation: new strategies call for new eco-indicators. Environmental Quality Management, 11: 51–61.
8. National Environmental Policy Plan 4 (2001). Nationaal Milieubeleidsplan 4: Een wereld en een wil. Ministry of Housing, Spatial Planning and the Environment, Den Haag (in Dutch).
9. WCED (1987). Our Common Future. Oxford University Press, Oxford.
10. Munasinghe M (1998). Climate change decision-making: science policy and economics. International Journal of Environment and Pollution, 10: 188–239.
11. Salomons W, Brils J (Eds) (2004). Contaminated Sediment in European River Basins. European Sediment Research Network, SedNet. TNO Den Helder, The Netherlands.
12. Owens PN, Batalla RJ, Collins AJ, Gomez B, Hicks DM, Horowitz AJ, Kondolf GM, Marden M, Page MJ, Peacock DH, Petticrew EL, Salomons W, Trustrum NA (2005). Fine-grained sediment in river systems: environmental significance and management issues. River Research and Applications, 21: 693–717.
13. Gibbons M, Limoges C, Nowotny H, Schwartzman S, Scott P, Trow M (1994). The New Production of Knowledge. Sage Publications, London.
14. Nowotny H, Scott P, Gibbons M (2003). 'Mode 2' revisited: the new production of knowledge. Minerva, 41: 179–194.
15. Ehrmann JR, Stinson BL (1999). Joint fact-finding and the use of technical expertise. In: Susskind L, McKearnan S, Thomas Larmer J (Eds), The Consensus Building Handbook: A Comprehensive Guide to Reaching Agreement. Sage Publications, London.
16. Eyquem J (2007). Using fluvial geomorphology to inform integrated river basin mangement. Water and Environment Journal, 21: 54–60.
17. Owens PN (2005). Conceptual models and budgets for sediment management at the river basin scale. Journal of Soils and Sediments, 5: 201–212.
18. Apitz S, White S (2003). A conceptual framework for river-basin-scale sediment management. Journal of Soils and Sediments, 3: 132–138.

19. Bergmann H, Maass V (2007). Sediment regulations and monitoring programmes in Europe. In: Heise S (Ed.), Sustainable Management of Sediment Resources: Sediment Risk Management and Communication. Elsevier, Amsterdam, pp. 207–231.
20. Parker A, Bergmann H, Heininger P, Leeks GJL, Old GH (2007). Sampling of sediment and suspended matter. In: Barceló D, Petrovic M (Eds), Sustainable Management of Sediment Resources: Sediment Quality and Impact Assessment of Pollutants. Elsevier, Amsterdam, pp. 1–34.
21. Hakstege AL (2007). Sediment management of nations of Europe: The Netherlands. In: Bortone G, Palumbo L (Eds), Sustainable Management of Sediment Resources: Sediment and Dredged Material Treatment. Elsevier, Amsterdam, pp. 11–21.
22. Detzner H-D (2007). Sediment management of nations of Europe: Germany. In: Bortone G, Palumbo L (Eds), Sustainable Management of Sediment Resources: Sediment and Dredged Material Treatment. Elsevier, Amsterdam, pp. 22–33.
23. Grosdemange D (2007). Sediment management of nations of Europe: France. In: Bortone G, Palumbo L (Eds), Sustainable Management of Sediment Resources: Sediment and Dredged Material Treatment. Elsevier, Amsterdam, pp. 40–45.
24. Netzband A, Reincke H, Bergmann H (2002). The River Elbe: a case study for the ecological and economical chain of sediments. Journal of Soils and Sediments, 2: 112–116.
25. Macleod CJA, Scholefield D, Haygarth PM (2007). Integration for sustainable catchment management. Science of the Total Environment, 373: 591–602.
26. Jones RJA, Le Bissonnais Y, Bazzoffi P, Díaz JS, Düwel O, Loj G, Øygarden L, Prasuhn V, Rydell B, Strauss P, Üveges JB, Vandekerckhove L, Yordanov Y (2004). Nature and extent of soil erosion in Europe. In: Van-Camp L, Bujarrabal B, Gentile A-R, Jones RJA, Montanarella L, Olazabal C, Selvaradjou S-K (Eds), Reports of the Technical Working Groups Established Under the Thematic Strategy for Soil Protection, EUR 21319 EN/1. Office for the Official Publications of the European Commission, Luxembourg, pp. 145–185.
27. Heise S, Förstner U, Westrich B, Jancke T, Karnahl J, Salomons W (2004). Inventory of historical contaminated sediment in Rhine basin and its tributaries. Report on Behalf of the Port of Rotterdam.
28. Heise S, Förstner U (2006). Risks from historical contaminated sediments in the Rhine basin. Water, Air and Soil Pollution: Focus, 6: 261–272.
29. Ellen GJ, Gerrits L, Slob AFL (2007). Risk perception and risk communication. In: Heise S (Ed.), Sustainable Management of Sediment Resources: Sediment Risk Management and Communication. Elsevier, Amsterdam, pp. 233–247.
30. Macklin MG, Brewer PA, Hudson-Edwards KA, Bird G, Coulthard TJ, Dennis IA, Lechler PJ, Miller JR, Turner JN (2006). A geomorphological approach to the management of rivers contaminated by metal mining. Geomorphology, 79: 423–447.
31. Fraser Basin Council (2007). http://www.fraserbasin.bc.ca
32. Apitz S, Carlon C, Oen A, White S (2007). Strategic frameworks for managing sediment risk at the basin and site-specific scale. In: Heise S (Ed.), Sustainable Management of Sediment Resources: Sediment Risk Management and Communication. Elsevier, Amsterdam, pp. 77–106.
33. Morgan RPC (2006). Managing sediment in the landscape: current practices and future vision. In: Owens PN, Collins AJ (Eds), Soil Erosion and Sediment Redistribution in River Catchments: Measurement, Modelling and Management. CABI, Wallingford, pp. 287–293.
34. ICPDR (2007). http://www.icpdr.org
35. Directive 2000/60/EC (2000). Of the European Parliament and of the Council of 23 October 2000 establishing a framework for Community action in the field of water policy (Water Framework Directive). Official Journal of the EC, L327.

36. ICPDR (2005). Danube Basin Analysis (WFD Roof Report 2004). Technical Report. International Commission for the Protection of the Danube River, Vienna, Austria.
37. ICPDR (2007). Management problems of sediment quality and quantity in the Danube River Basin. Issue Paper (in preparation). International Commission for the Protection of the Danube River, Vienna, Austria.
38. SedNet (2007). Sediment Management – An Essential Element of River Basin Management Plans. SedNet, The Netherlands, http://www.sednet.org
39. UNESCO-ISI (2007). http://www.itces.org/isi/
40. AquaTerra (2007). http://www.attempto-projects.de/aquaterra
41. Slob AFL, Rijnveld M, Chapman AS, Strosser P (2007). Challenges of linking scientific knowledge to river basin management policy: AquaTerra as a case study. Environmental Pollution, accepted for publication.
42. Waals JFM (2001). CO_2-Reduction in Housing: Experiences in Building and Urban Renewal Projects in the Netherlands. Rozenberg Publishers, Amsterdam.
43. Susskind L, McKearnan S, Thomas Larmer J (Eds) (1999). The Consensus Building Handbook: A Comprehensive Guide to Reaching Agreement. Sage Publications, London.
44. Rivershed Society of British Columbia (2007). http://www.rivershed.com
45. MELP (1994). Stream Stewardship: A Guide for Planners and Developers. Ministry of Environment, Lands and Parks, Vancouver.
46. RiskBase (2007). http://www.riskbase.info
47. SedNet (2007). Sediment Research Needs. SedNet document presented to the European Commission, http://www.sednet.org

Index